Student Edition

Eureka Math
Grade 3
Module 3

Special thanks go to the Gordon A. Cain Center and to the Department of Mathematics at Louisiana State University for their support in the development of *Eureka Math* .

For a free *Eureka Math* Teacher Resource Pack, Parent Tip Sheets, and more please visit www.Eureka.tools

Printed in the U.S.A.

This book may be purchased from the publisher at eureka-math.org

7 8 9 10 LSC 24 23 22 21 20

ISBN 978-1-63255-299-0

Name _____   Date _____

1. a. Solve. Shade in the multiplication facts that you already know. Then, shade in the facts for sixes, sevens, eights, and nines that you can solve using the commutative property.

×	1	2	3	4	5	6	7	8	9	10
1		2	3							
2		4		8				16		
3						18				
4					20					
5										50
6		12								
7										
8										
9										
10										

b. Complete the chart. Each bag contains 7 apples.

Number of Bags	2		4	5	
Total Number of Apples		21			42

2. Use the array to write two different multiplication sentences.

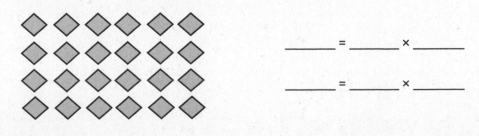

_____ = _____ × _____

_____ = _____ × _____

3. Complete the equations.

a. 2 sevens = _____ twos

       = ___14___

b. 3 _____ = 6 threes

       = _____

c. 10 eights = 8 _____

       = _____

d. 4 × _____ = 6 × 4

       = _____

e. 8 × 5 = _____ × 8

       = _____

f. _____ × 7 = 7 × _____

       = ___28___

g. 3 × 9 = 10 threes − _____ three

       = _____

h. 10 fours − 1 four = _____ × 4

       = _____

i. 8 × 4 = 5 fours + _____ fours

       = _____

j. _____ fives + 1 five = 6 × 5

       = _____

k. 5 threes + 2 threes = _____ × _____

       = _____

l. _____ twos + _____ twos = 10 twos

       = _____

Name _____   Date _____

1. Complete the charts below.

   a. A tricycle has 3 wheels.

Number of Tricycles	3		5		7
Total Number of Wheels		12		18	

   b. A tiger has 4 legs.

Number of Tigers			7	8	9
Total Number of Legs	20	24			

   c. A package has 5 erasers.

Number of Packages	6				10
Total Number of Erasers		35	40	45	

2. Write two multiplication facts for each array.

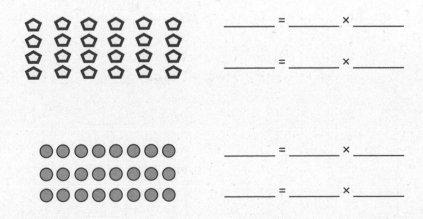

_____ = _____ × _____

_____ = _____ × _____

_____ = _____ × _____

_____ = _____ × _____

3. Match the expressions.

3×6 7 threes

3 sevens 2×10

2 eights 9×5

5×9 8×2

10 twos 6×3

---

4. Complete the equations.

a.   2 sixes = _____ twos

       = __12__

b.   _____ × 6 = 6 threes

       = _____

c. 4×8 = _____ × 4

       = _____

d. $4 \times$ _____ = _____ × 4

       = __28__

e.   5 twos + 2 twos = _____ × _____

       = _____

f.   _____ fives + 1 five = 6×5

       = _____

Lesson 1:   Study commutativity to find known facts of 6, 7, 8, and 9.

EUREKA MATH

Name _____   Date _____

1. Each 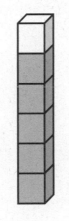 has a value of 7.

Unit form: 5 _____

Facts: 5 × _____ = _____ × 5

Total = _____

Unit form: 6 sevens = _____ sevens + _____ seven

= 35 + _____

= _____

Facts: _____ × _____ = _____

_____ × _____ = _____

2. a. Each dot has a value of 8

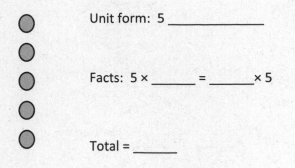

Unit form: 5 _____

Facts: 5 × _____ = _____ × 5

Total = _____

b. Use the fact above to find 8 × 6. Show your work using pictures, numbers, or words.

Lesson 2:   Apply the distributive and commutative properties to relate
                  multiplication facts 5 × n + n to 6 × n and n × 6 where n is the size
                  of the unit.

©2015 Great Minds. eureka-math.org
G3-M3-SE-B2-1.3.1-01.2016

3. An author writes 9 pages of her book each week. How many pages does she write in 7 weeks?
    Use a fives fact to solve.

4. Mrs. Gonzalez buys a total of 32 crayons for her classroom. Each pack contains 8 crayons. How many
    packs of crayons does Mrs. Gonzalez buy?

5. Hannah has $500. She buys a camera for $435 and 4 other items for $9 each. Now Hannah wants to buy
    speakers for $50. Does she have enough money to buy the speakers? Explain.

EUREKA
MATH™

Lesson 2:   Apply the distributive and commutative properties to relate
             multiplication facts 5 × n + n to 6 × n and n × 6 where n is the size
             of the unit.
             ©2015 Great Minds. eureka-math.org
             G3-M3-SE-B2-1.3.1-01.2016

7

Name _____   Date _____

1. Each 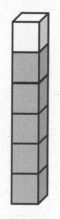 has a value of 9.

Unit form: _____

Facts: 5 × _____ = _____ × 5

Total = _____

Unit form: 6 nines = _____ nines + _____ nine

 = 45 + _____

= _____

Facts: _____ × _____ = _____

_____ × _____ = _____

Lesson 2:   Apply the distributive and commutative properties to relate
            multiplication facts 5 × n + n to 6 × n and n × 6 where n is the size
            of the unit.

©2015 Great Minds. eureka-math.org
G3-M3-SE-B2-1.3.1-01.2016

2. There are 6 blades on each windmill. How many total blades are on 7 windmills? Use a fives fact to solve.

3. Juanita organizes her magazines into 3 equal piles. She has a total of 18 magazines. How many magazines are in each pile?

4. Markuo spends $27 on some plants. Each plant costs $9. How many plants does he buy?

This page intentionally left blank

3. Miss Potts used a total of 28 cups of flour to bake some bread. She used 4 cups of flour for each loaf of bread. How many loaves of bread did she bake? Represent the problem using multiplication and division sentences and a letter for the unknown. Then, solve the problem.

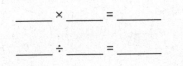

4. At a table tennis tournament, two games went on for a total of 32 minutes. One game took 12 minutes longer than the other. How long did it take to complete each game? Use letters to represent the unknowns. Solve the problem.

CHALLENGE!

EUREKA
MATH™

Lesson 3:   Multiply and divide with familiar facts using a letter to represent
                 the unknown.

©2015 Great Minds. eureka-math.org
G3-M3-SE-B2-1.3.1-01.2016

13

Name _____   Date _____

1. a. Complete the pattern.

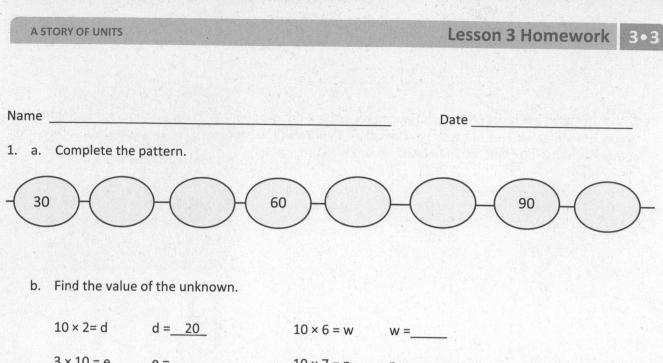

30   60   90

   b. Find the value of the unknown.

   $10 \times 2 = d$ $d = \underline{\ 20\ }$ $10 \times 6 = w$ $w = \underline{\hspace{1cm}}$

   $3 \times 10 = e$ $e = \underline{\hspace{1cm}}$ $10 \times 7 = n$ $n = \underline{\hspace{1cm}}$

   $f = 4 \times 10$ $f = \underline{\hspace{1cm}}$ $g = 8 \times 10$ $g = \underline{\hspace{1cm}}$

   $p = 5 \times 10$ $p = \underline{\hspace{1cm}}$

2. Each equation contains a letter representing the unknown. Find the value of the unknown.

$8 \div 2 = n$	$n = \underline{\hspace{1cm}}$
$3 \times a = 12$	$a = \underline{\hspace{1cm}}$
$p \times 8 = 40$	$p = \underline{\hspace{1cm}}$
$18 \div 6 = c$	$c = \underline{\hspace{1cm}}$
$d \times 4 = 24$	$d = \underline{\hspace{1cm}}$
$h \div 7 = 5$	$h = \underline{\hspace{1cm}}$
$6 \times 3 = f$	$f = \underline{\hspace{1cm}}$
$32 \div y = 4$	$y = \underline{\hspace{1cm}}$

Lesson 3:   Multiply and divide with familiar facts using a letter to represent
                        the unknown.

EUREKA MATH

3. Pedro buys 4 books at the fair for $7 each.

    a. What is the total amount Pedro spends on 4 books? Use the letter *b* to represent the total amount Pedro spends, and then solve the problem.

    b. Pedro hands the cashier 3 ten dollar bills. How much change will he receive? Write an equation to solve. Use the letter *c* to represent the unknown.

4. On field day, the first-grade dash is 25 meters long. The third-grade dash is twice the distance of the first-grade dash. How long is the third-grade dash? Use a letter to represent the unknown and solve.

This page intentionally left blank

Name _____   Date _____

1. Skip-count by six to fill in the blanks. Match each number in the count-by with its multiplication fact.

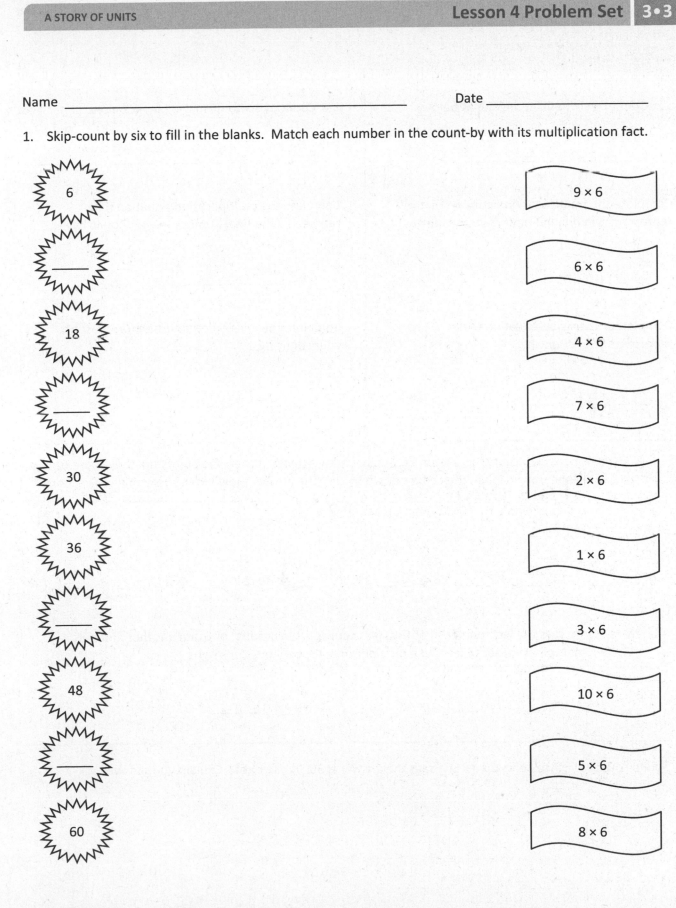

6

____

18

____

30

36

____

48

____

60

9 × 6

6 × 6

4 × 6

7 × 6

2 × 6

1 × 6

3 × 6

10 × 6

5 × 6

8 × 6

EUREKA MATH™

Lesson 4:   Count by units of 6 to multiply and divide using number bonds
             to decompose.

17

2. Count by six to fill in the blanks below.

    6, _____, _____, _____

    Complete the multiplication equation that represents the final number in your count-by.

    6 × _____ = _____

    Complete the division equation that represents your count-by.

    _____ ÷ 6 = _____

3. Count by six to fill in the blanks below.

    6, _____, _____, _____, _____, _____, _____

    Complete the multiplication equation that represents the final number in your count-by.

    6 × _____ = _____

    Complete the division equation that represents your count-by.

    _____ ÷ 6 = _____

4. Mrs. Byrne's class skip-counts by six for a group counting activity. When she points up, they count up by six, and when she points down, they count down by six. The arrows show when she changes direction.

    a. Fill in the blanks below to show the group counting answers.

    ↑ 0, 6, _____, 18, _____ ↓ _____, 12 ↑ _____, 24, 30, _____ ↓ 30, 24, _____ ↑ 24, _____, 36, _____, 48

    b. Mrs. Byrne says the last number that the class counts is the product of 6 and another number. Write a multiplication sentence and a division sentence to show she's right.

    6 × _____ = 48   48 ÷ 6 = _____

5. Julie counts by six to solve 6 × 7. She says the answer is 36. Is she right? Explain your answer.

Lesson 4:   Count by units of 6 to multiply and divide using number bonds to decompose.

©2015 Great Minds. eureka-math.org
G3-M3-SE-B2-1.3.1-01.2016

Name _____   Date _____

1. Use number bonds to help you skip-count by six by either making a ten or adding to the ones.

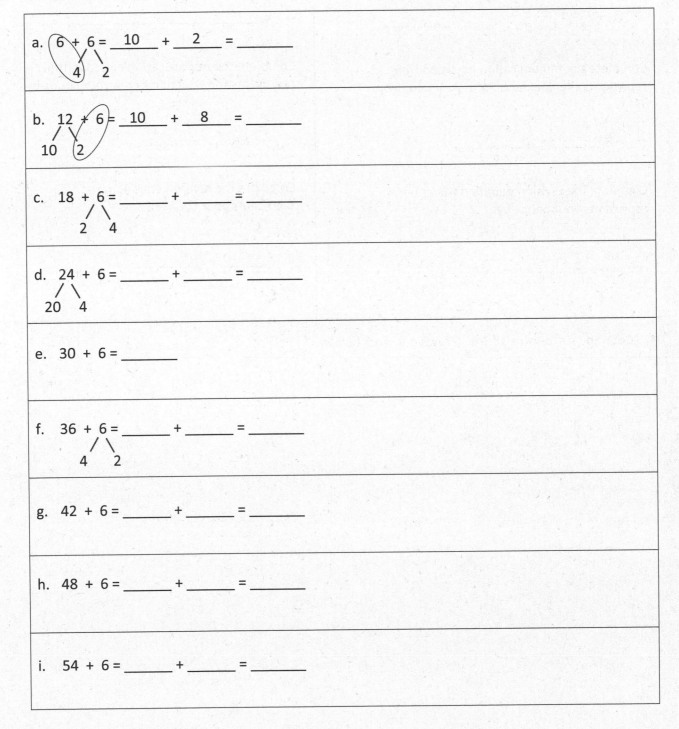

a. 6 + 6 = ___10___ + ___2___ = _____
       4   2

b. 12 + 6 = ___10___ + ___8___ = _____
    10   2

c. 18 + 6 = _____ + _____ = _____
       2   4

d. 24 + 6 = _____ + _____ = _____
    20   4

e. 30 + 6 = _____

f. 36 + 6 = _____ + _____ = _____
       4   2

g. 42 + 6 = _____ + _____ = _____

h. 48 + 6 = _____ + _____ = _____

i. 54 + 6 = _____ + _____ = _____

EUREKA MATH

Lesson 4:   Count by units of 6 to multiply and divide using number bonds to decompose.

19

2. Count by six to fill in the blanks below.

6, _____, _____, _____, _____

Complete the multiplication equation that represents the final number in your count-by.

6 × _____ = _____

Complete the division equation that represents your count-by.

_____ ÷ 6 = _____

3. Count by six to fill in the blanks below.

6, _____, _____, _____, _____, _____

Complete the multiplication equation that represents the final number in your count-by.

6 × _____ = _____

Complete the division equation that represents your count-by.

_____ ÷ 6 = _____

4. Count by six to solve 48 ÷ 6. Show your work below.

Lesson 4: Count by units of 6 to multiply and divide using number bonds to decompose.

Name _____   Date _____

1. Skip-count by seven to fill in the blanks in the fish bowls. Match each count-by to its multiplication expression. Then, use the multiplication expression to write the related division fact directly to the right.

7

7 × 6   _____ ÷ 7 = _____

_____

7 × 3   _____ ÷ 7 = _____

21

7 × 8   _____ ÷ 7 = _____

_____

7 × 7   _____ ÷ 7 = _____

_____

7 × 1   _____ ÷ 7 = _____

42

7 × 10   _____ ÷ 7 = _____

49

7 × 9   _____ ÷ 7 = _____

_____

7 × 4   _____ ÷ 7 = _____

_____

7 × 2   _____ ÷ 7 = _____

_____

7 × 5   _____ ÷ 7 = _____

Lesson 5:   Count by units of 7 to multiply and divide using number bonds to decompose.

21

©2015 Great Minds. eureka-math.org
G3-M3-SE-B2-1.3.1-01.2016

2. Complete the count-by seven sequence below. Then, write a multiplication equation and a division equation to represent each blank you filled in.

7, 14, _____, 28, _____, 42, _____, _____, 63, _____

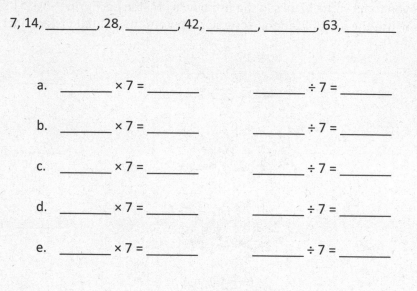

a. _____ × 7 = _____   _____ ÷ 7 = _____

b. _____ × 7 = _____   _____ ÷ 7 = _____

c. _____ × 7 = _____   _____ ÷ 7 = _____

d. _____ × 7 = _____   _____ ÷ 7 = _____

e. _____ × 7 = _____   _____ ÷ 7 = _____

3. Abe says 3 × 7 = 21 because 1 seven is 7, 2 sevens are 14, and 3 sevens are 14 + 6 + 1, which equals 21. Why did Abe add 6 and 1 to 14 when he is counting by seven?

4. Molly says she can count by seven 6 times to solve 7 × 6. James says he can count by six 7 times to solve this problem. Who is right? Explain your answer.

Lesson 5:   Count by units of 7 to multiply and divide using number bonds to decompose.

©2015 Great Minds. eureka-math.org
G3-M3-SE-B2-1.3.1-01.2016

EUREKA MATH™

Name _____   Date _____

1. Use number bonds to help you skip-count by seven by making ten or adding to the ones.

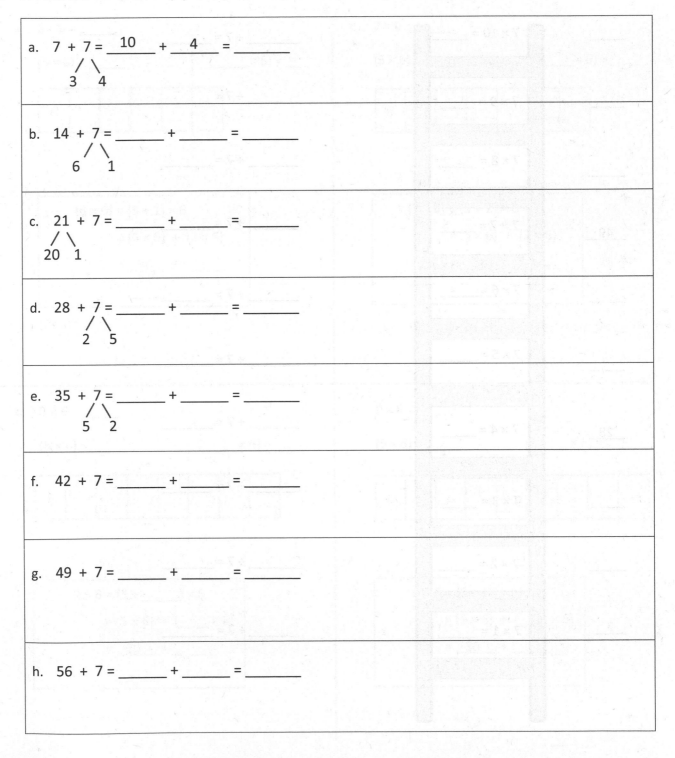

a. 7 + 7 = ___10___ + ___4___ = _____
         / \
        3   4

b. 14 + 7 = _____ + _____ = _____
          / \
         6   1

c. 21 + 7 = _____ + _____ = _____
        / \
       20 1

d. 28 + 7 = _____ + _____ = _____
          / \
         2   5

e. 35 + 7 = _____ + _____ = _____
          / \
         5   2

f. 42 + 7 = _____ + _____ = _____

g. 49 + 7 = _____ + _____ = _____

h. 56 + 7 = _____ + _____ = _____

 EUREKA MATH™   Lesson 5:   Count by units of 7 to multiply and divide using number bonds   23
                                       to decompose.

©2015 Great Minds. eureka-math.org
G3-M3-SE-B2-1.3.1-01.2016

2. Break apart 54 to solve 54 ÷ 6.

54 ÷ 6

30 ÷ 6   24 ÷ 6

54 ÷ 6 = (30 ÷ 6) + (_____ ÷ 6)

= 5 + _____

= _____

3. Break apart 49 to solve 49 ÷ 7.

49 ÷ 7

35 ÷ 7

49 ÷ 7 = (35 ÷ 7) + (_____ ÷ 7)

= 5 + _____

= _____

4. Robert says that he can solve 6 × 8 by thinking of it as (5 × 8) + 8. Is he right? Draw a picture to help explain your answer.

5. Kelly solves 42 ÷ 7 by using a number bond to break apart 42 into two parts. Show what her work might look like below.

Lesson 6:   Use the distributive property as a strategy to multiply and divide using units of 6 and 7.

©2015 Great Minds. eureka-math.org
G3-M3-SE-B2-1.3.1-01.2016

Name _____   Date _____

1. Label the tape diagrams. Then, fill in the blanks below to make the statements true.

a. **6 × 7**= _____

$(5 \times 7) =$ ____ $(\_\_\_ \times 7) =$ ____

7					

$(6 \times 7) = (5 + 1) \times 7$

$= (5 \times 7) + (1 \times 7)$

$= \underline{\quad 35 \quad} + \underline{\qquad}$

$= \underline{\qquad}$

b. **7 × 7 =** _____

$(5 \times 7) =$ ____ $(\_\_\_ \times 7) =$ ____

7						

$(7 \times 7) = (5 + 2) \times 7$

$= (5 \times 7) + (2 \times 7)$

$= \underline{\quad 35 \quad} + \underline{\qquad}$

$= \underline{\qquad}$

c. **8 × 7 =** _____

$(5 \times 7) =$ ____ $(\_\_\_ \times 7) =$ ____

7							

$8 \times 7 = (5 + \underline{\quad\quad}) \times 7$

$= (5 \times 7) + (\_\_\_ \times 7)$

$= \underline{\quad 35 \quad} + \underline{\qquad}$

$= \underline{\qquad}$

d. **9 × 7 =** _____

$(5 \times 7) =$ ____ $(\_\_\_ \times 7) =$ ____

7								

$9 \times 7 = (5 + \underline{\quad\quad}) \times 7$

$= (5 \times 7) + (\_\_\_ \times 7)$

$= \underline{\quad 35 \quad} + \underline{\qquad}$

$= \underline{\qquad}$

2. Break apart 54 to solve 54 ÷ 6.

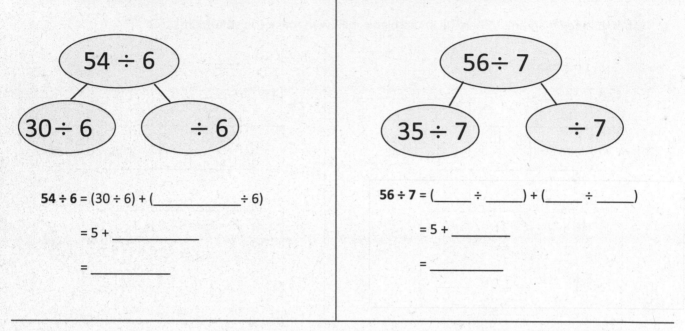

    54 ÷ 6 = (30 ÷ 6) + (_____ ÷ 6)

         = 5 + _____

         = _____

3. Break apart 56 to solve 56 ÷ 7

    56 ÷ 7 = (_____ ÷ _____) + (_____ ÷ _____)

         = 5 + _____

         = _____

4. Forty-two third grade students sit in 6 equal rows in the auditorium. How many students sit in each row? Show your thinking.

5. Ronaldo solves 7 × 6 by thinking of it as (5 × 7) + 7. Is he correct? Explain Ronaldo's strategy.

Lesson 6:   Use the distributive property as a strategy to multiply and divide
using units of 6 and 7.

©2015 Great Minds. eureka-math.org
G3-M3-SE-B2-1.3.1-01.2016

Name _____   Date _____

1. Match the words to the correct equation.

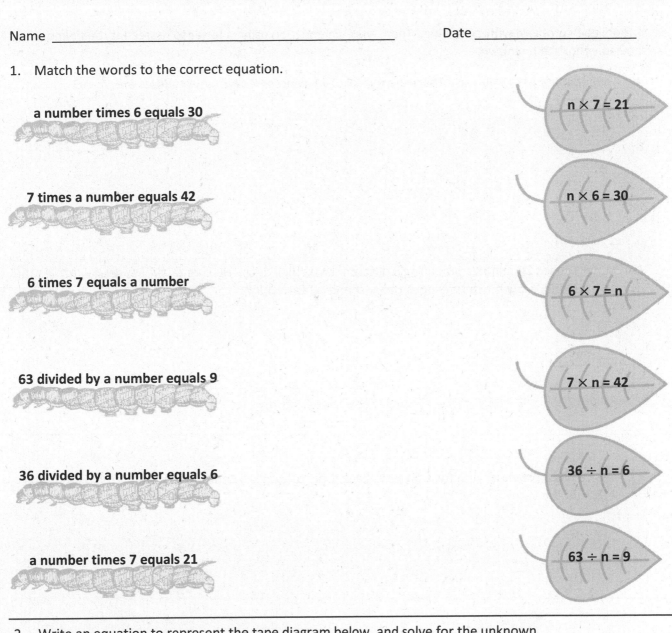

a number times 6 equals 30

7 times a number equals 42

6 times 7 equals a number

63 divided by a number equals 9

36 divided by a number equals 6

a number times 7 equals 21

$n \times 7 = 21$

$n \times 6 = 30$

$6 \times 7 = n$

$7 \times n = 42$

$36 \div n = 6$

$63 \div n = 9$

2. Write an equation to represent the tape diagram below, and solve for the unknown.

8	8	8	8	8	8

k

Equation: _____

Lesson 7:   Interpret the unknown in multiplication and division to model and solve problems using units of 6 and 7.

29

©2015 Great Minds. eureka-math.org
G3-M3-SE-B2-1.3.1-01.2016

3. Model each problem with a drawing. Then, write an equation using a letter to represent the unknown, and solve for the unknown.

a. Each student gets 3 pencils. There are a total of 21 pencils. How many students are there?

b. Henry spends 24 minutes practicing 6 different basketball drills. He spends the same amount of time on each drill. How much time does Henry spend on each drill?

c. Jessica has 8 pieces of yarn for a project. Each piece of yarn is 6 centimeters long. What is the total length of the yarn?

d. Ginny measures 6 milliliters of water into each beaker. She pours a total of 54 milliliters. How many beakers does Ginny use?

Lesson 7:   Interpret the unknown in multiplication and division to model and
             solve problems using units of 6 and 7.

©2015 Great Minds. eureka-math.org
G3-M3-SE-B2-1.3.1-01.2016

Name _____ Date _____

1. Match the words on the arrow to the correct equation on the target.

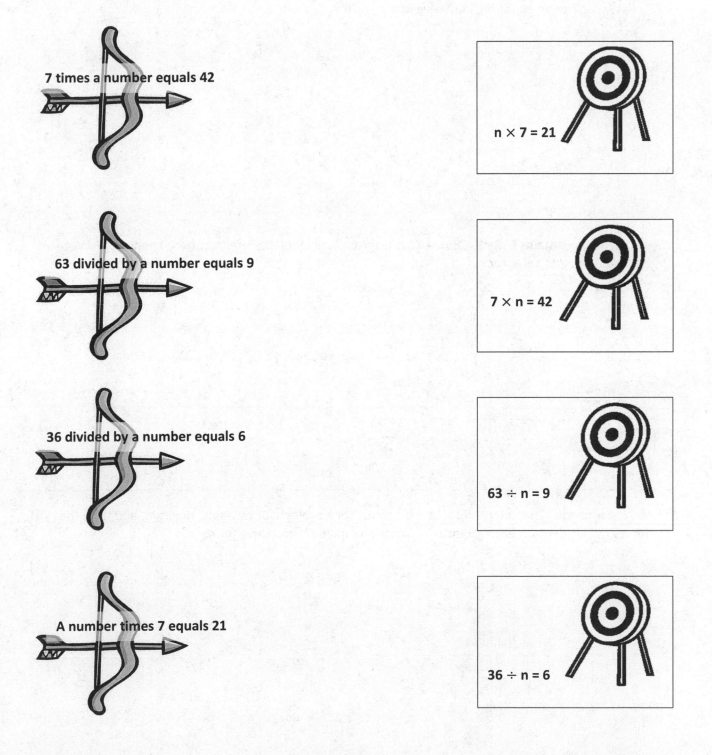

7 times a number equals 42

63 divided by a number equals 9

36 divided by a number equals 6

A number times 7 equals 21

$n \times 7 = 21$

$7 \times n = 42$

$63 \div n = 9$

$36 \div n = 6$

EUREKA
MATH™

Lesson 7: Interpret the unknown in multiplication and division to model and
solve problems using units of 6 and 7.

31

2. Ari sells 6 boxes of pens at the school store.

   a. Each box of pens sells for $7. Draw a tape diagram, and label the total amount of money he makes as m. Write an equation, and solve for m.

   b. Each box contains 6 pens. Draw a tape diagram, and label the total number of pens as p. Write an equation, and solve for p.

3. Mr. Lucas divides 28 students into 7 equal groups for a project. Draw a tape diagram, and label the number of students in each group as n. Write an equation, and solve for n.

Lesson 7:   Interpret the unknown in multiplication and division to model and solve problems using units of 6 and 7.

©2015 Great Minds. eureka-math.org
G3-M3-SE-B2-1.3.1-01.2016

Name _____   Date _____

1. Solve.

   a. $(12 - 4) + 6 =$ _____

   b. $12 - (4 + 6) =$ _____

   c. _____ $= 15 - (7 + 3)$

   d. _____ $= (15 - 7) + 3$

   e. _____ $= (3 + 2) \times 6$

   f. _____ $= 3 + (2 \times 6)$

   g. $4 \times (7 - 2) =$ _____

   h. $(4 \times 7) - 2 =$ _____

   i. _____ $= (12 \div 2) + 4$

   j. _____ $= 12 \div (2 + 4)$

   k. $9 + (15 \div 3) =$ _____

   l. $(9 + 15) \div 3 =$ _____

   m. $60 \div (10 - 4) =$ _____

   n. $(60 \div 10) - 4 =$ _____

   o. _____ $= 35 + (10 \div 5)$

   p. _____ $= (35 + 10) \div 5$

2. Use parentheses to make the equations true.

a. $16 - 4 + 7 = 19$	b. $16 - 4 + 7 = 5$
c. $2 = 22 - 15 + 5$	d. $12 = 22 - 15 + 5$
e. $3 + 7 \times 6 = 60$	f. $3 + 7 \times 6 = 45$
g. $5 = 10 \div 10 \times 5$	h. $50 = 100 \div 10 \times 5$
i. $26 - 5 \div 7 = 3$	j. $36 = 4 \times 25 - 16$

©2015 Great Minds. eureka-math.org
G3-M3-SE-B2-1.3.1-01.2016

3. The teacher writes 24 ÷ 4 + 2 = _____ on the board. Chad says it equals 8. Samir says it equals 4.
    Explain how placing the parentheses in the equation can make both answers true.

4. Natasha solves the equation below by finding the sum of 5 and 12. Place the parentheses in the equation
    to show her thinking. Then, solve.

    12 + 15 ÷ 3 = _____

5. Find two possible answers to the expression 7 + 3 × 2 by placing the parentheses in different places.

Name _____   Date _____

1. Solve.

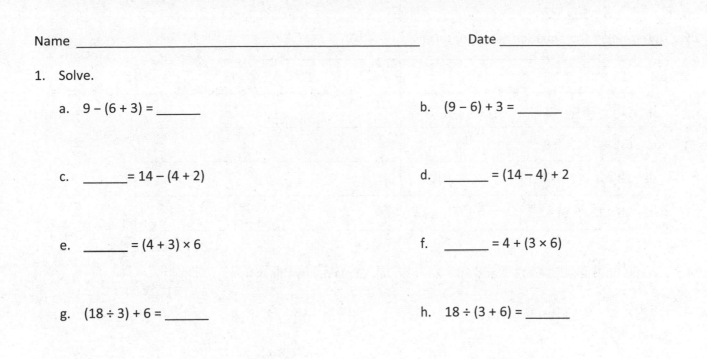

    a.   9 – (6 + 3) = _____

    b.   (9 – 6) + 3 = _____

    c.   _____ = 14 – (4 + 2)

    d.   _____ = (14 – 4) + 2

    e.   _____ = (4 + 3) × 6

    f.   _____ = 4 + (3 × 6)

    g.   (18 ÷ 3) + 6 = _____

    h.   18 ÷ (3 + 6) = _____

2. Use parentheses to make the equations true.

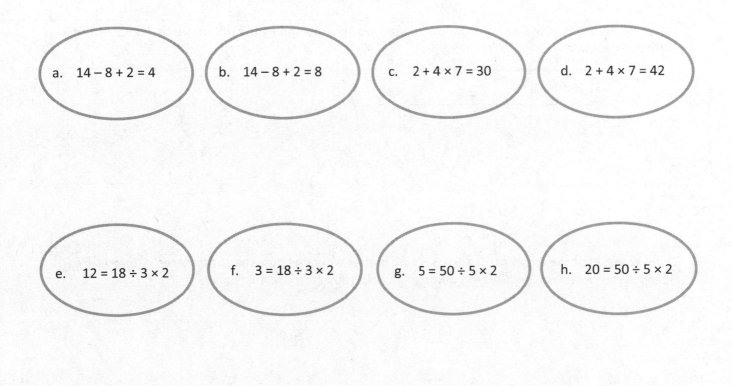

    a.   14 – 8 + 2 = 4

    b.   14 – 8 + 2 = 8

    c.   2 + 4 × 7 = 30

    d.   2 + 4 × 7 = 42

    e.   12 = 18 ÷ 3 × 2

    f.   3 = 18 ÷ 3 × 2

    g.   5 = 50 ÷ 5 × 2

    h.   20 = 50 ÷ 5 × 2

©2015 Great Minds. eureka-math.org
G3-M3-SE-B2-1.3.1-01.2016

3. Determine if the equation is true or false.

a. $(15 - 3) \div 2 = 6$	*Example:* True
b. $(10 - 7) \times 6 = 18$	
c. $(35 - 7) \div 4 = 8$	
d. $28 = 4 \times (20 - 13)$	
e. $35 = (22 - 8) \div 5$	

4. Jerome finds that $(3 \times 6) \div 2$ and $18 \div 2$ are equal. Explain why this is true.

5. Place parentheses in the equation below so that you solve by finding the difference between 28 and 3. Write the answer.

$$4 \times 7 - 3 = \underline{\hspace{2cm}}$$

6. Johnny says that the answer to $2 \times 6 \div 3$ is 4 no matter where he puts the parentheses. Do you agree? Place parentheses around different numbers to help you explain his thinking.

            Lesson 8:   Understand the function of parentheses and apply to solving problems.

©2015 Great Minds. eureka-math.org
G3-M3-SE-B2-1.3.1-01.2016

Name _____ Date _____

Solve the following pairs of problems. Circle the pairs where both problems have the same answer.

1. a. $7 + (6 + 4)$

    b. $(7 + 6) + 4$

5. a. $(3 + 2) \times 5$

    b. $3 + (2 \times 5)$

2. a. $(3 \times 2) \times 4$

    b. $3 \times (2 \times 4)$

6. a. $(8 \div 2) \times 2$

    b. $8 \div (2 \times 2)$

3. a. $(2 \times 1) \times 5$

    b. $2 \times (1 \times 5)$

7. a. $(9 - 5) + 3$

    b. $9 - (5 + 3)$

4. a. $(4 \times 2) \times 2$

    b. $4 \times (2 \times 2)$

8. a. $(8 \times 5) - 4$

    b. $8 \times (5 - 4)$

EUREKA
MATH™

Lesson 9:   Model the associative property as a strategy to multiply.

37

©2015 Great Minds. eureka-math.org
G3-M3-SE-B2-1.3.1-01.2016

Name _____ Date _____

1. Use the array to complete the equation.

a. 3 × 12 = _____

b. (3 × 3) × 4

    = _____ × 4

    = _____

c. 3 × 14 = _____

d. (_____ × _____) × 7

    = _____ × _____

    = _____

Lesson 9:   Model the associative property as a strategy to multiply.

2. Place parentheses in the equations to simplify. Then, solve. The first one has been done for you.

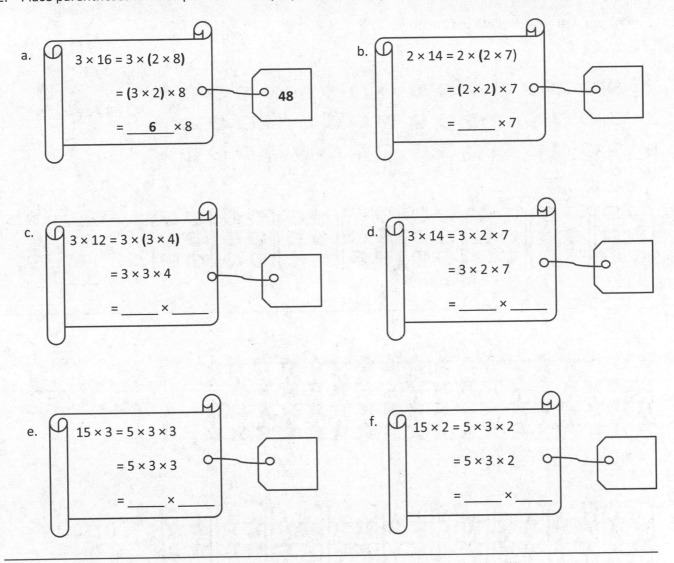

a.
$3 \times 16 = 3 \times (2 \times 8)$

$= (3 \times 2) \times 8$ 48

$= \underline{\quad 6 \quad} \times 8$

b.
$2 \times 14 = 2 \times (2 \times 7)$

$= (2 \times 2) \times 7$

$= \underline{\quad\quad} \times 7$

c.
$3 \times 12 = 3 \times (3 \times 4)$

$= 3 \times 3 \times 4$

$= \underline{\quad\quad} \times \underline{\quad\quad}$

d.
$3 \times 14 = 3 \times 2 \times 7$

$= 3 \times 2 \times 7$

$= \underline{\quad\quad} \times \underline{\quad\quad}$

e.
$15 \times 3 = 5 \times 3 \times 3$

$= 5 \times 3 \times 3$

$= \underline{\quad\quad} \times \underline{\quad\quad}$

f.
$15 \times 2 = 5 \times 3 \times 2$

$= 5 \times 3 \times 2$

$= \underline{\quad\quad} \times \underline{\quad\quad}$

3. Charlotte finds the answer to 16×2 by thinking about 8×4. Explain her strategy.

Name _____ Date _____

1. Use the array to complete the equation.

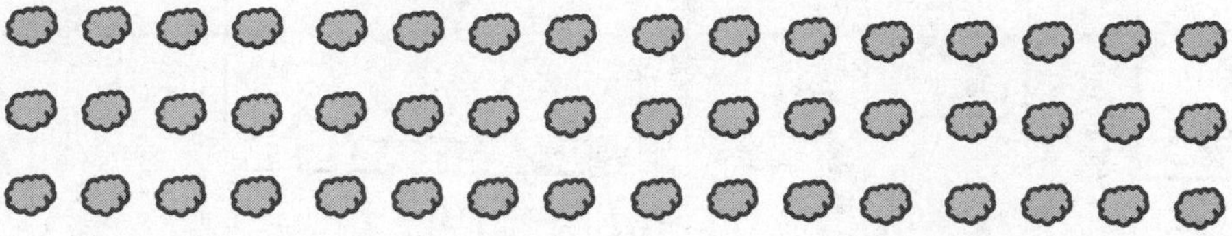

a. $3 \times 16 =$ _____

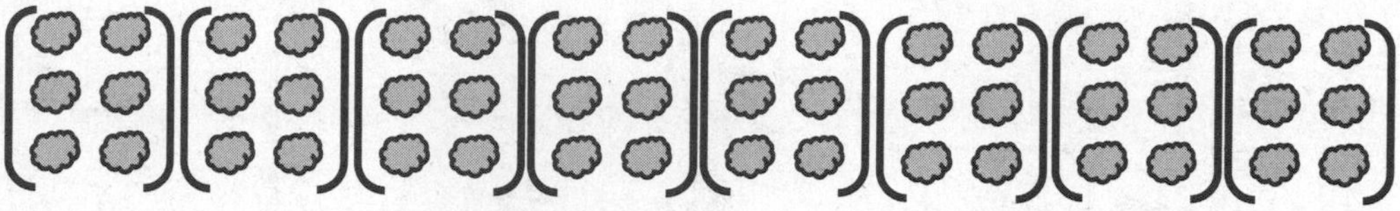

b. $(3 \times$ _____$) \times 8$

   $=$ _____ $\times$ _____

   $=$ _____

c. $4 \times 18 =$ _____

d. $(4 \times$ _____$) \times 9$

   $=$ _____ $\times$ _____

   $=$ _____

Lesson 9: Model the associative property as a strategy to multiply.

EUREKA MATH

2. Place parentheses in the equations to simplify and solve.

$12 \times 4 = (6 \times 2) \times 4$

$= 6 \times (2 \times 4)$ $= \underline{\textbf{48}}$

$= 6 \times \underline{\textbf{8}}$

a. $3 \times 14 = 3 \times (2 \times 7)$

$= 3 \times 2 \times 7$ $= \underline{\hspace{1cm}}$

$= \underline{\hspace{1cm}} \times 7$

b. $3 \times 12 = 3 \times (3 \times 4)$

$= 3 \times 3 \times 4$ $= \underline{\hspace{1cm}}$

$= \underline{\hspace{1cm}} \times 4$

3. Solve. Then, match the related facts.

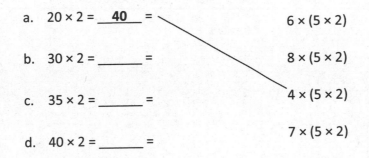

a. $20 \times 2 = \underline{\textbf{40}} =$ $6 \times (5 \times 2)$

b. $30 \times 2 = \underline{\hspace{1cm}} =$ $8 \times (5 \times 2)$

c. $35 \times 2 = \underline{\hspace{1cm}} =$ $4 \times (5 \times 2)$

d. $40 \times 2 = \underline{\hspace{1cm}} =$ $7 \times (5 \times 2)$

EUREKA MATH™

Lesson 9: Model the associative property as a strategy to multiply.

41

©2015 Great Minds. eureka-math.org
G3-M3-SE-B2-1.3.1-01.2016

This page intentionally left blank

Name _____   Date _____

1. Label the arrays. Then, fill in the blanks below to make the statements true.

a. $8 \times 8 =$ _____

    $(8 \times 5) =$ _____ $(8 \times$ _____ $) =$ _____

    $8 \times 8 = 8 \times (5 +$ _____ $)$

          $= (8 \times 5) + (8 \times$ _____ $)$

          $=$ _40_ $+$ _____

          $=$ _____

b. $8 \times 9 = 9 \times 8 =$ _____

    $(8 \times 5) =$ _____ $(8 \times$ _____ $) =$ _____

    $9 \times 8 = 8 \times (5 +$ _____ $)$

          $= (8 \times 5) + (8 \times$ _____ $)$

          $=$ _40_ $+$ _____

          $=$ _____

2. Break apart and distribute to solve $56 \div 8$.

           $56 \div 8$

    $40 \div 8$ $16 \div 8$

    $56 \div 8 = (40 \div 8) + ($ _____ $\div 8)$

       $= 5 +$ _____

       $=$ _____

3. Break apart and distribute to solve $72 \div 8$.

           $72 \div 8$

    $40 \div 8$

    $72 \div 8 = (40 \div 8) + ($ _____ $\div 8)$

       $= 5 +$ _____

       $=$ _____

4. An octagon has 8 sides. Skip-count to find the total number of sides on 9 octagons.

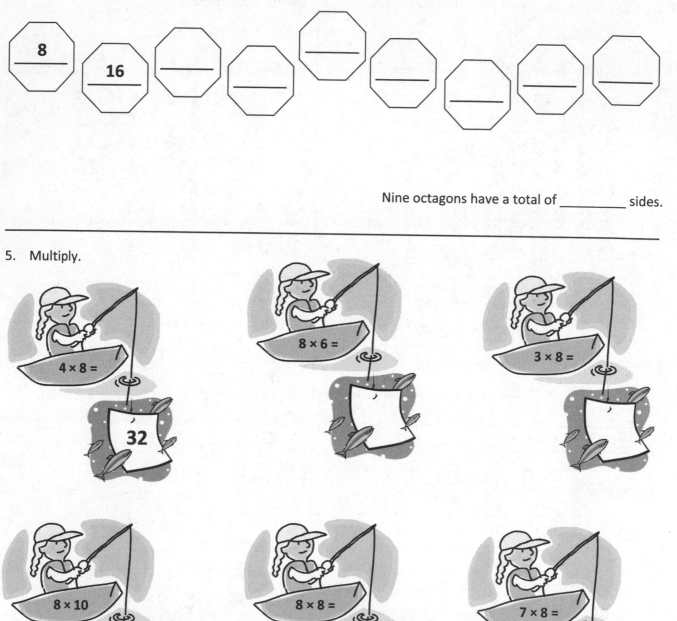

8

16

___

___

___

___

___

___

___

Nine octagons have a total of _____ sides.

5. Multiply.

$4 \times 8 =$

32

8 × 6 =

$3 \times 8 =$

8 × 10

8 × 8 =

$7 \times 8 =$

Lesson 10:   Use the distributive property as a strategy to multiply and divide.

6. Match.

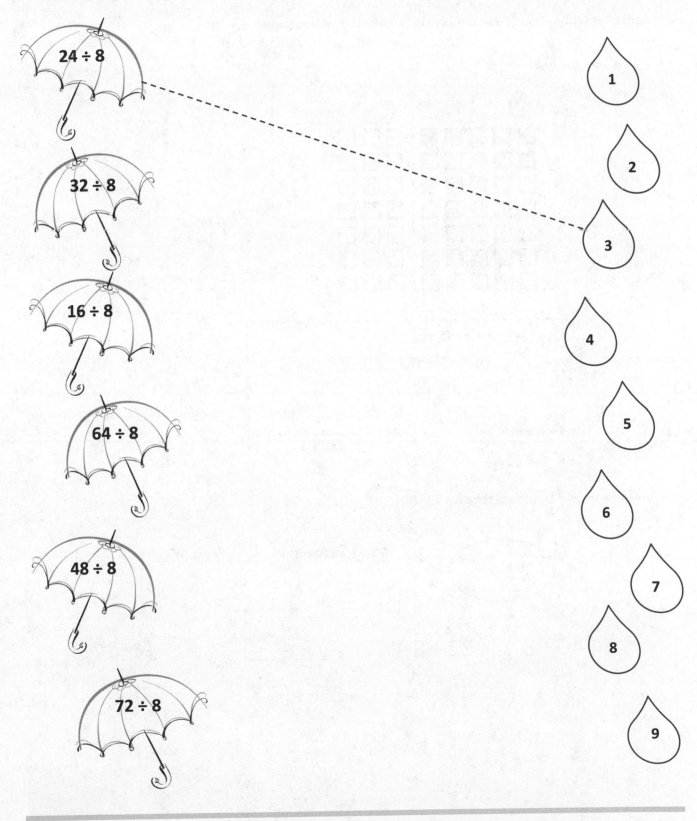

Lesson 10:   Use the distributive property as a strategy to multiply and divide.

45

Lesson 10 Homework 3•3

Name _____ Date _____

1. Label the array. Then, fill in the blanks to make the statements true.

$8 \times 7 = 7 \times 8 = $_____

(7 × 5) = _____   (7 × _____) = _____

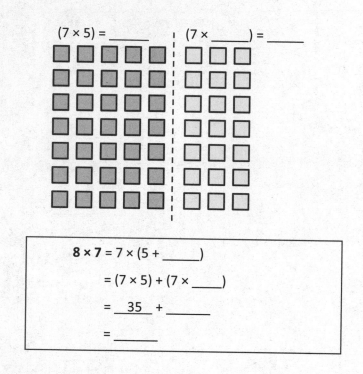

$8 \times 7 = 7 \times (5 + $_____$)$

$= (7 \times 5) + (7 \times $_____$)$

$= $__35__$ + $_____

$= $_____

2. Break apart and distribute to solve 72 ÷ 8.

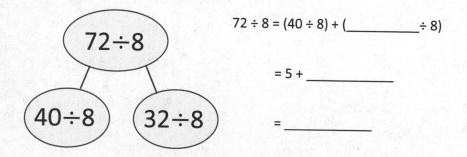

$72 \div 8 = (40 \div 8) + ($_____$\div 8)$

$= 5 + $_____

$= $_____

©2015 Great Minds. eureka-math.org
G3-M3-SE-B2-1.3.1-01.2016

3. Count by 8. Then, match each multiplication problem with its value.

   ___8___ , _____ , _____ , _____ , _____ , _____ , _____ , _____ , _____ , _____

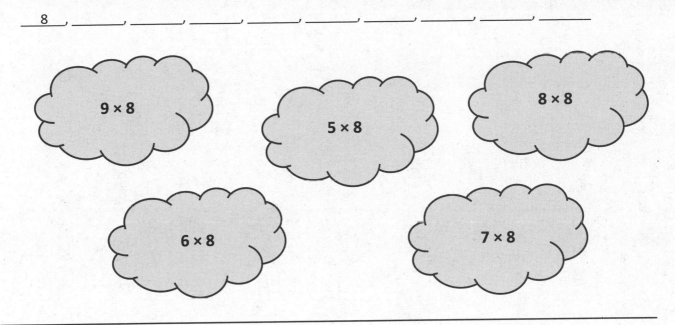

4. Divide.

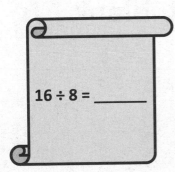

16 ÷ 8 = _____

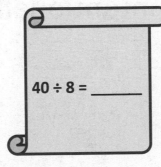

40 ÷ 8 = _____

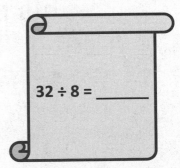

32 ÷ 8 = _____

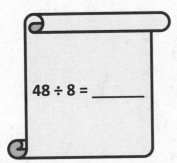

48 ÷ 8 = _____

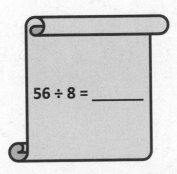

56 ÷ 8 = _____

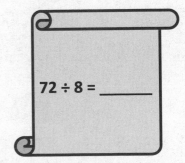

72 ÷ 8 = _____

This page intentionally left blank

Name _____   Date _____

1. Ms. Santor divides 32 students into 8 equal groups for a field trip. Draw a tape diagram, and label the number of students in each group as *n*. Write an equation, and solve for *n*.

2. Tara buys 6 packs of printer paper. Each pack of paper costs $8. Draw a tape diagram, and label the total amount she spends as *m*. Write an equation, and solve for *m*.

3. Mr. Reed spends $24 on coffee beans. How many kilograms of coffee beans does he buy? Draw a tape diagram, and label the total amount of coffee beans he buys as *c*. Write an equation, and solve for *c*.

$8 for 1 kg

Lesson 11:   Interpret the unknown in multiplication and division to model and solve problems.

49

©2015 Great Minds. eureka-math.org
G3-M3-SE-B2-1.3.1-01.2016

4. Eight boys equally share 4 packs of baseball cards. Each pack contains 10 cards. How many cards does each boy get?

5. There are 8 bags of yellow and green balloons. Each bag contains 7 balloons. If there are 35 yellow balloons, how many green balloons are there?

6. The fruit seller packs 72 oranges into bags of 8 each. He sells all the oranges at $4 a bag. How much money did he receive?

Lesson 11: Interpret the unknown in multiplication and division to model and solve problems.

©2015 Great Minds. eureka-math.org
G3-M3-SE-B2-1.3.1-01.2016

Name _____ Date _____

1. Jenny bakes 10 cookies. She puts 7 chocolate chips on each cookie. Draw a tape diagram, and label the total amount of chocolate chips as *c*. Write an equation, and solve for *c*.

2. Mr. Lopez arranges 48 dry erase markers into 8 equal groups for his math stations. Draw a tape diagram, and label the number of dry erase markers in each group as *v*. Write an equation, and solve for *v*.

3. There are 35 computers in the lab. Five students each turn off an equal number of computers. How many computers does each student turn off? Label the unknown as *m*, and then solve.

Lesson 11:   Interpret the unknown in multiplication and division to model and
              solve problems.

©2015 Great Minds. eureka-math.org
G3-M3-SE-B2-1.3.1-01.2016

51

4. There are 9 bins of books. Each bin has 6 comic books. How many comic books are there altogether?

5. There are 8 trail mix bags in one box. Clarissa buys 5 boxes. She gives an equal number of bags of trail mix to 4 friends. How many bags of trail mix does each friend receive?

6. Leo earns $8 each week for doing chores. After 7 weeks, he buys a gift and has $38 left. How much money does he spend on the gift?

Lesson 11:   Interpret the unknown in multiplication and division to model and solve problems.

©2015 Great Minds. eureka-math.org
G3-M3-SE-B2-1.3.1-01.2016

. Find the total value of the shaded blocks.

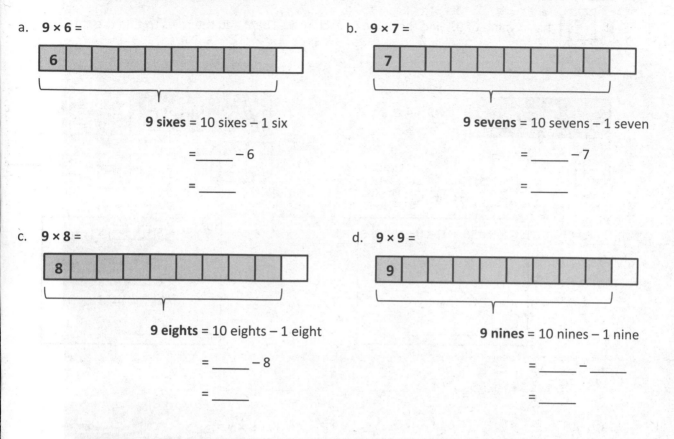

a. **9 × 6 =**

6

9 sixes = 10 sixes − 1 six

= _____ − 6

= _____

b. **9 × 7 =**

7

9 sevens = 10 sevens − 1 seven

= _____ − 7

= _____

c. **9 × 8 =**

8

9 eights = 10 eights − 1 eight

= _____ − 8

= _____

d. **9 × 9 =**

9

9 nines = 10 nines − 1 nine

= _____ − _____

= _____

. Matt buys a pack of postage stamps. He counts 9 rows of 4 stamps. He thinks of 10 fours to find the total number of stamps. Show the strategy that Matt might have used to find the total number of stamps.

Lesson 12: Apply the distributive property and the fact 9 = 10 − 1 as a strategy to multiply.

©2015 Great Minds. eureka-math.org
G3-M3-SE-B2-1.3.1-01.2016

Name _____ Date _____

1. Each 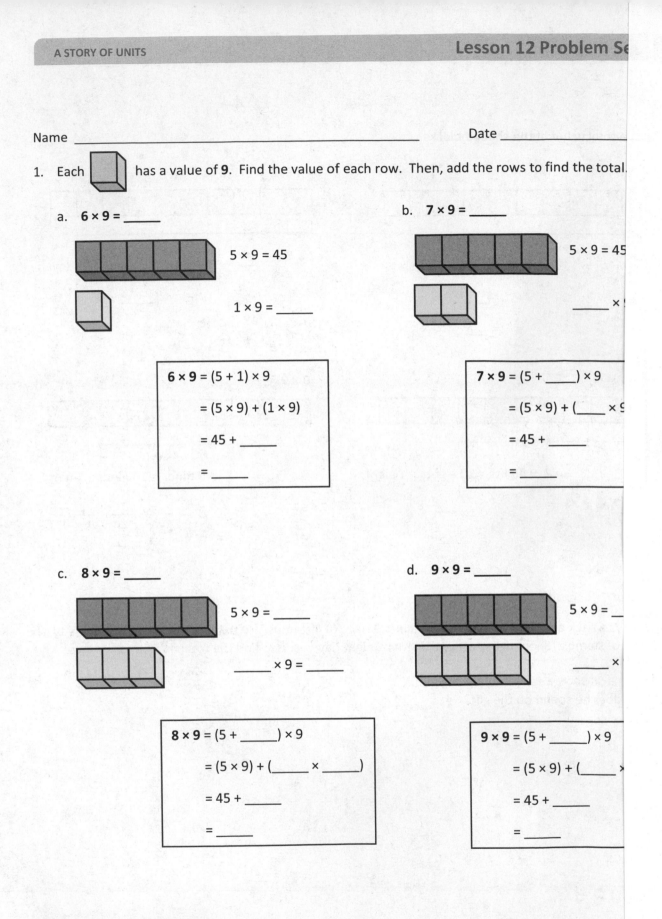 has a value of **9**. Find the value of each row. Then, add the rows to find the total.

a. **6 × 9 =** _____

$5 \times 9 = 45$

$1 \times 9 =$ _____

$\begin{aligned} \mathbf{6 \times 9} &= (5 + 1) \times 9 \\ &= (5 \times 9) + (1 \times 9) \\ &= 45 + \underline{\hspace{1cm}} \\ &= \underline{\hspace{1cm}} \end{aligned}$

b. **7 × 9 =** _____

$5 \times 9 = 45$

_____ × !

$\begin{aligned} \mathbf{7 \times 9} &= (5 + \underline{\hspace{1cm}}) \times 9 \\ &= (5 \times 9) + (\underline{\hspace{0.5cm}} \times \text{9} \\ &= 45 + \underline{\hspace{1cm}} \\ &= \underline{\hspace{1cm}} \end{aligned}$

c. **8 × 9 =** _____

$5 \times 9 =$ _____

_____ × 9 = _____

$\begin{aligned} \mathbf{8 \times 9} &= (5 + \underline{\hspace{1cm}}) \times 9 \\ &= (5 \times 9) + (\underline{\hspace{1cm}} \times \underline{\hspace{1cm}}) \\ &= 45 + \underline{\hspace{1cm}} \\ &= \underline{\hspace{1cm}} \end{aligned}$

d. **9 × 9 =** _____

$5 \times 9 =$ _____

_____ ×

$\begin{aligned} \mathbf{9 \times 9} &= (5 + \underline{\hspace{1cm}}) \times 9 \\ &= (5 \times 9) + (\underline{\hspace{1cm}} \times \\ &= 45 + \underline{\hspace{1cm}} \\ &= \underline{\hspace{1cm}} \end{aligned}$

3

Lesson 12:   Apply the distributive property and the fact 9 = 10 − 1 as a strategy
to multiply.

©2015 Great Minds. eureka-math.org
G3-M3-SE-B2-1.3.1-01.2016

5

4. Match.

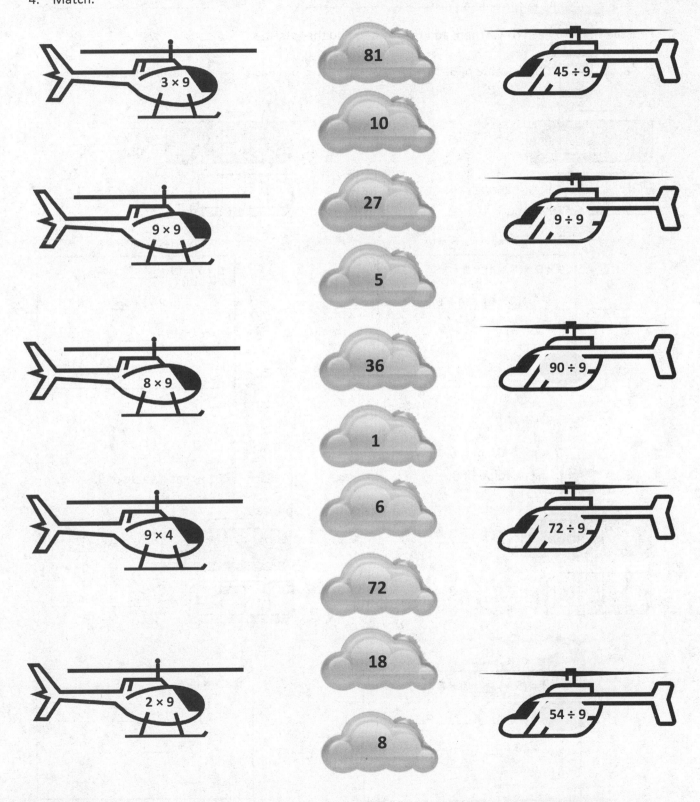

EUREKA MATH

Lesson 12:   Apply the distributive property and the fact 9 = 10 − 1 as a strategy
to multiply.

©2015 Great Minds. eureka-math.org
G3-M3-SE-B2-1.3.1-01.2016

55

Name _____   Date _____

1. Find the value of each row. Then, add the rows to find the total.

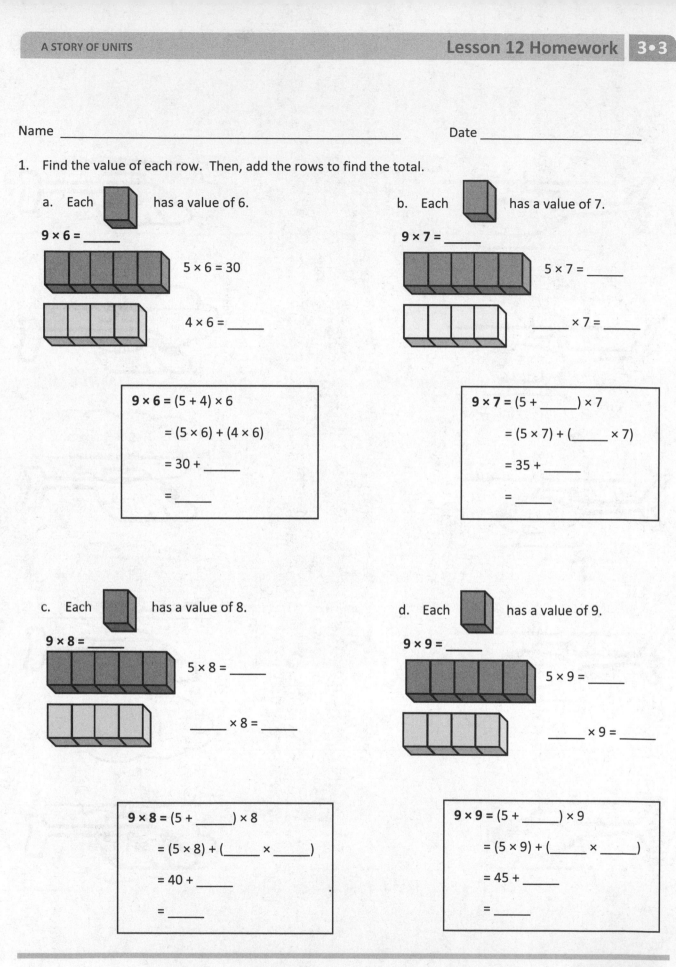

a. Each ▨ has a value of 6.

9 × 6 = _____

5 × 6 = 30

4 × 6 = _____

9 × 6 = (5 + 4) × 6

    = (5 × 6) + (4 × 6)

    = 30 + _____

    = _____

b. Each ▨ has a value of 7.

9 × 7 = _____

5 × 7 = _____

_____ × 7 = _____

9 × 7 = (5 + _____) × 7

    = (5 × 7) + (_____ × 7)

    = 35 + _____

    = _____

c. Each ▨ has a value of 8.

9 × 8 = _____

5 × 8 = _____

_____ × 8 = _____

9 × 8 = (5 + _____) × 8

    = (5 × 8) + (_____ × _____)

    = 40 + _____

    = _____

d. Each ▨ has a value of 9.

9 × 9 = _____

5 × 9 = _____

_____ × 9 = _____

9 × 9 = (5 + _____) × 9

    = (5 × 9) + (_____ × _____)

    = 45 + _____

    = _____

Lesson 12:   Apply the distributive property and the fact 9 = 10 − 1 as a strategy
              to multiply.

©2015 Great Minds. eureka-math.org
G3-M3-SE-B2-1.3.1-01.2016

This page intentionally left blank

2. Match.

a. **9 fives** = 10 fives − 1 five

= 50 − 5 --------------

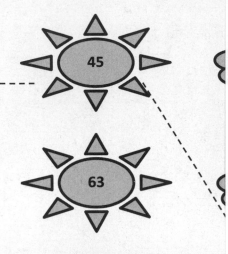

b. **9 sixes** = 10 sixes − 1 six

=_____ − 6

c. **9 sevens** = 10 sevens − 1 seven

= _____ − 7

d. **9 eights** = 10 eights − 1 eight

= _____ − 8

e. **9 nines** = 10 nines − 1 nine

= _____ − _____

f. **9 fours** = 10 fours − 1 four

= _____ − _____

Lesson 12:   Apply the distributive property and the fact 9 = 10 − 1 as
to multiply.

©2015 Great Minds. eureka-math.org
G3-M3-SE-B2-1.3.1-01.2016

58

tape diagram

Lesson 12: Apply the distributive property and the fact 9 = 10 − 1 as a strategy
to multiply.

59

©2015 Great Minds. eureka-math.org
G3-M3-SE-B2-1.3.1-01.2016

This page intentionally left blank

Name _____   Date _____

1. a. Skip-count by nine.

    ____9____, _____, _____, ___36___, _____, _____, _____, ___72___, _____, _____

    b. Look at the *tens* place in the count-by. What is the pattern?

    c. Look at the *ones* place in the count-by. What is the pattern?

2. Complete to make true statements.

    a. 10 more than 0 is ___10___,   f. 10 more than 45 is _____,
        1 less is ___9___.   1 less is _____.
        $1 \times 9 =$ ___9___ $6 \times 9 =$ _____

    b. 10 more than 9 is ___19___,   g. 10 more than 54 is _____,
        1 less is ___18___.   1 less is _____.
        $2 \times 9 =$ _____ $7 \times 9 =$ _____

    c. 10 more than 18 is _____,   h. 10 more than 63 is _____,
        1 less is _____.   1 less is _____.
        $3 \times 9 =$ _____ $8 \times 9 =$ _____

    d. 10 more than 27 is _____,   i. 10 more than 72 is _____,
        1 less is _____.   1 less is _____.
        $4 \times 9 =$ _____ $9 \times 9 =$ _____

    e. 10 more than 36 is _____,   j. 10 more than 81 is _____,
        1 less is _____.   1 less is _____.
        $5 \times 9 =$ _____ $10 \times 9 =$ _____

3. a. Analyze the equations in Problem 2. What is the pattern?

    b. Use the pattern to find the next 4 facts. Show your work.

        11 × 9 =   12 × 9 =   13 × 9 =   14 × 9 =

    c. Kent notices another pattern in Problem 2. His work is shown below. He sees the following:
        ▪   The tens digit in the product is 1 less than the number of groups.
        ▪   The ones digit in the product is 10 minus the number of groups.

	Tens digit	Ones digit
2 × 9 = <u>18</u> →	<u>1</u> = 2 − 1	<u>8</u> = 10 − 2
3 × 9 = <u>27</u> →	<u>2</u> = 3 − 1	<u>7</u> = 10 − 3
4 × 9 = <u>36</u> →	<u>3</u> = 4 − 1	<u>6</u> = 10 − 4
5 × 9 = <u>45</u> →	<u>4</u> = 5 − 1	<u>5</u> = 10 − 5

        Use Kent's strategy to solve 6 × 9 and 7 × 9.

    d. Show an example of when Kent's pattern doesn't work.

©2015 Great Minds. eureka-math.org
G3-M3-SE-B2-1.3.1-01.2016

4. Each equation contains a letter representing the unknown. Find the value of each unknown. Then, write the letters that match the answers to solve the riddle.

$a \times 9 = 54$

$a =$ _____

$81 \div 9 = g$

$g =$ _____

$9 \times d = 72$

$d =$ _____

$e \times 9 = 63$

$e =$ _____

$o \div 9 = 10$

$o =$ _____

$9 \times n = 27$

$n =$ _____

$t \times 9 = 18$

$t =$ _____

$9 \times s = 36$

$s =$ _____

$i \div 9 = 5$

$i =$ _____

How do you make one vanish?

___ ___ ___ ___ " ___ " ___ ___ ___ ___ , ___ ___ ___ ___ ___ !
6   8   8   6   9   6   3   8   45 2   4   9   90 3   7

EUREKA
MATH™

Name _____   Date _____

1. a. Skip-count by nines down from 90.

   ___90___ , _____ , __72__ , _____ , _____ , _____ , __36__ , _____ , _____ , _____

   b. Look at the *tens* place in the count-by. What is the pattern?

   c. Look at the *ones* place in the count-by. What is the pattern?

2. Each equation contains a letter representing the unknown. Find the value of each unknown.

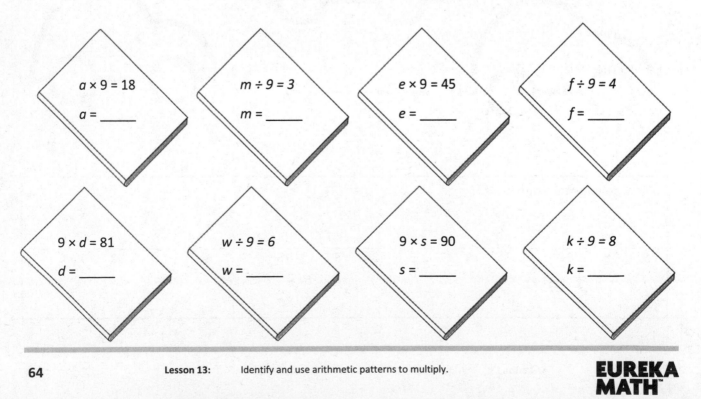

$a \times 9 = 18$

$a =$ _____

$m \div 9 = 3$

$m =$ _____

$e \times 9 = 45$

$e =$ _____

$f \div 9 = 4$

$f =$ _____

$9 \times d = 81$

$d =$ _____

$w \div 9 = 6$

$w =$ _____

$9 \times s = 90$

$s =$ _____

$k \div 9 = 8$

$k =$ _____

EUREKA
MATH

3. Solve.

a. What is 10 more than 0? _____
   What is 1 less? _____
   $1 \times 9 =$ _____

b. What is 10 more than 9? _____
   What is 1 less? _____
   $2 \times 9 =$ _____

c. What is 10 more than 18? _____
   What is 1 less? _____
   $3 \times 9 =$ _____

d. What is 10 more than 27? _____
   What is 1 less? _____
   $4 \times 9 =$ _____

e. What is 10 more than 36? _____
   What is 1 less? _____
   $5 \times 9 =$ _____

f. What is 10 more than 45? _____
   What is 1 less? _____
   $6 \times 9 =$ _____

g. What is 10 more than 54? _____
   What is 1 less? _____
   $7 \times 9 =$ _____

h. What is 10 more than 63? _____
   What is 1 less? _____
   $8 \times 9 =$ _____

i. What is 10 more than 72? _____
   What is 1 less? _____
   $9 \times 9 =$ _____

j. What is 10 more than 81? _____
   What is 1 less? _____
   $10 \times 9 =$ _____

4. Explain the pattern in Problem 3, and use the pattern to solve the next 3 facts.

$11 \times 9 =$ _____ $12 \times 9 =$ _____ $13 \times 9 =$ _____

This page intentionally left blank

Name _____ Date _____

1. a. Multiply. Then, add the tens digit and ones digit of each product.

    1 × 9 = 9   0 + 9 =   9

    2 × 9 = 18   1 + 8 = _____

    3 × 9 = ____   ____ + ____ = _____

    4 × 9 = ____   ____ + ____ = _____

    5 × 9 = ____   ____ + ____ = _____

    6 × 9 = ____   ____ + ____ = _____

    7 × 9 = ____   ____ + ____ = _____

    8 × 9 = ____   ____ + ____ = _____

    9 × 9 = ____   ____ + ____ = _____

    10 × 9 = ____   ____ + ____ = _____

    b. What is the sum of the digits in each product? How can this strategy help you check your work with the nines facts?

    c. Araceli continues to count by nines. She writes, "90, 99, 108, 117, 126, 135, 144, 153, 162, 171, 180, 189, 198. Wow! The sum of the digits is still 9." Is she correct? Why or why not?

2. Araceli uses the number of groups in 8 × 9 to help her find the product. She uses 8 − 1 = 7 to get the digit in the tens place and 10 − 8 = 2 to get the digit in the ones place. Use her strategy to find 4 more facts.

3. Dennis calculates 9 × 8 by thinking about it as 80 − 8 = 72. Explain Dennis' strategy.

4. Sonya figures out the answer to 7 × 9 by putting down her right index finger (shown). What is the answer? Explain how to use Sonya's finger strategy.

Name _____   Date _____

1. a. Multiply. Then, add the digits in each product.

$10 \times 9 = 90$	_9_ + _0_ = _9_
$9 \times 9 = 81$	_8_ + _1_ = _9_
$8 \times 9 =$	____ + ____ = ____
$7 \times 9 =$	____ + ____ = ____
$6 \times 9 =$	____ + ____ = ____
$5 \times 9 =$	____ + ____ = ____
$4 \times 9 =$	____ + ____ = ____
$3 \times 9 =$	____ + ____ = ____
$2 \times 9 =$	____ + ____ = ____
$1 \times 9 =$	____ + ____ = ____

b. What pattern did you notice in Problem 1(a)? How can this strategy help you check your work with nines facts?

2. Thomas calculates 9 × 7 by thinking about it as 70 − 7 = 63. Explain Thomas' strategy.

---

3. Alexia figures out the answer to 6 × 9 by lowering the thumb on her right hand (shown). What is the answer? Explain Alexia's strategy.

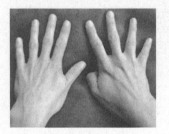

---

4. Travis writes 72 = 9 × 8. Is he correct? Explain at least 2 strategies Travis can use to check his work.

Name _____ Date _____

Write an equation, and use a letter to represent the unknown for Problems 1–6.

1. Mrs. Parson gave each of her grandchildren $9. She gave a total of $36. How many grandchildren does Mrs. Parson have?

2. Shiva pours 27 liters of water equally into 9 containers.   How many liters of water are in each container?

3. Derek cuts 7 pieces of wire. Each piece is 9 meters long. What is the total length of the 7 pieces?

Lesson 15:   Interpret the unknown in multiplication and division to model and solve problems.

71

©2015 Great Minds. eureka-math.org
G3-M3-SE-B2-1.3.1-01.2016

4. Aunt Deena and Uncle Chris share the cost of a limousine ride with their 7 friends. The ride cost a total of $63. If everyone shares the cost equally, how much does each person pay?

5. Cara bought 9 packs of beads. There are 10 beads in each pack. She always uses 30 beads to make each necklace. How many necklaces can she make if she uses all the beads?

6. There are 8 erasers in a set. Damon buys 9 sets. After giving some erasers away, Damon has 35 erasers left. How many erasers did he give away?

Lesson 15:   Interpret the unknown in multiplication and division to model and solve problems.

©2015 Great Minds. eureka-math.org
G3-M3-SE-B2-1.3.1-01.2016

Name _____   Date _____

1. The store clerk equally divides 36 apples among 9 baskets. Draw a tape diagram, and label the number of apples in each basket as *a*. Write an equation, and solve for *a*.

2. Elijah gives each of his friends a pack of 9 almonds. He gives away a total of 45 almonds. How many packs of almonds did he give away? Model using a letter to represent the unknown, and then solve.

3. Denice buys 7 movies. Each movie costs $9. What is the total cost of 7 movies? Use a letter to represent the unknown. Solve.

4. Mr. Doyle shares 1 roll of bulletin board paper equally with 8 teachers. The total length of the roll is 72 meters. How much bulletin board paper does each teacher get?

5. There are 9 pens in a pack. Ms. Ochoa buys 9 packs. After giving her students some pens, she has 27 pens left. How many pens did she give away?

6. Allen buys 9 packs of trading cards. There are 10 cards in each pack. He can trade 30 cards for a comic book. How many comic books can he get if he trades all of his cards?

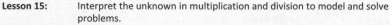

Lesson 15: Interpret the unknown in multiplication and division to model and solve problems.

EUREKA MATH

Name _____   Date _____

1. Complete.

   a. _____ × 1 = 6   b. _____ ÷ 7 = 0   c. 8 × _____ = 8   d. 9 ÷ _____ = 9

   e. 0 ÷ 5 = _____   f. _____ × 0 = 0   g. 4 ÷ _____ = 1   h. _____ × 1 = 3

2. Match each equation with its solution.

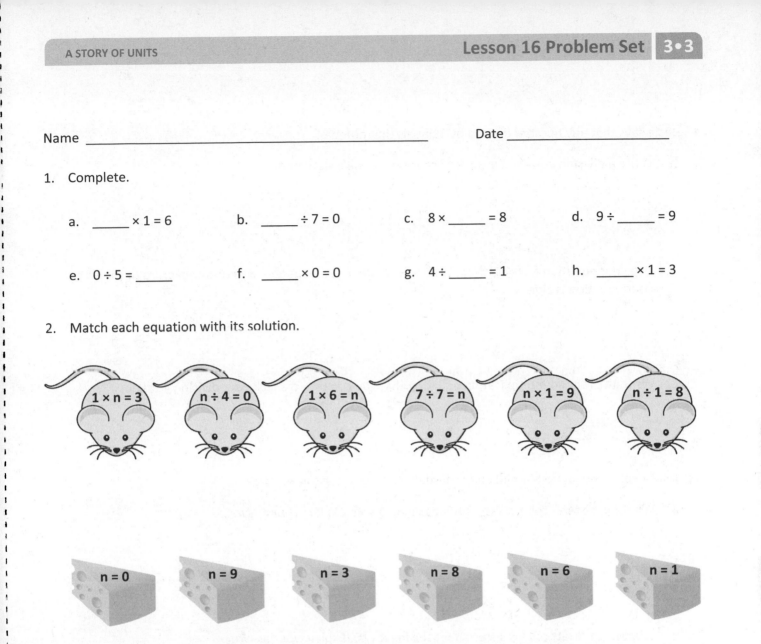

3. Let *n* be a number. Complete the blanks below with the products.

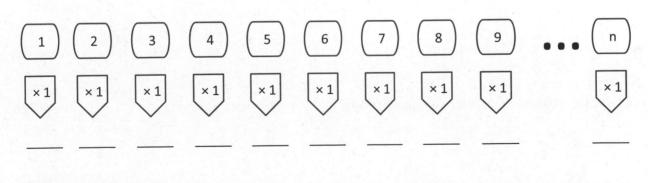

   What pattern do you notice?

Lesson 16:   Reason about and explain arithmetic patterns using units
of 0 and 1 as they relate to multiplication and division.

75

©2015 Great Minds. eureka-math.org
G3-M3-SE-B2-1.3.1-01.2016

4. Josie says that any number divided by 1 equals that number.

   a. Write a division equation using n to represent Josie's statement.

   b. Use your equation from Part (a). Let $n = 6$. Write a new equation, and draw a picture to show that your equation is true.

   c. Write the related multiplication equation that you can use to check your division equation.

5. Matt explains what he learned about dividing with zero to his little sister.

   a. What might Matt tell his sister about solving $0 \div 9$? Explain your answer.

   b. What might Matt tell his sister about solving $8 \div 0$? Explain your answer.

   c. What might Matt tell his sister about solving $0 \div 0$? Explain your answer.

Lesson 16:   Reason about and explain arithmetic patterns using units of 0 and 1 as they relate to multiplication and division.

©2015 Great Minds. eureka-math.org
G3-M3-SE-B2-1.3.1-01.2016

Name _____   Date _____

1. Complete.

a. $4 \times 1 = $ _____   b. $4 \times 0 = $ _____   c. _____ $\times 1 = 5$ d. _____ $\div 5 = 0$

e. $6 \times $ _____ $= 6$ f. _____ $\div 6 = 0$ g. $0 \div 7 = $ _____   h. $7 \times $ _____ $= 0$

i. $8 \div $ _____ $= 8$ j. _____ $\times 8 = 8$ k. $9 \times $ _____ $= 9$ l. $9 \div $ _____ $= 1$

2. Match each equation with its solution.

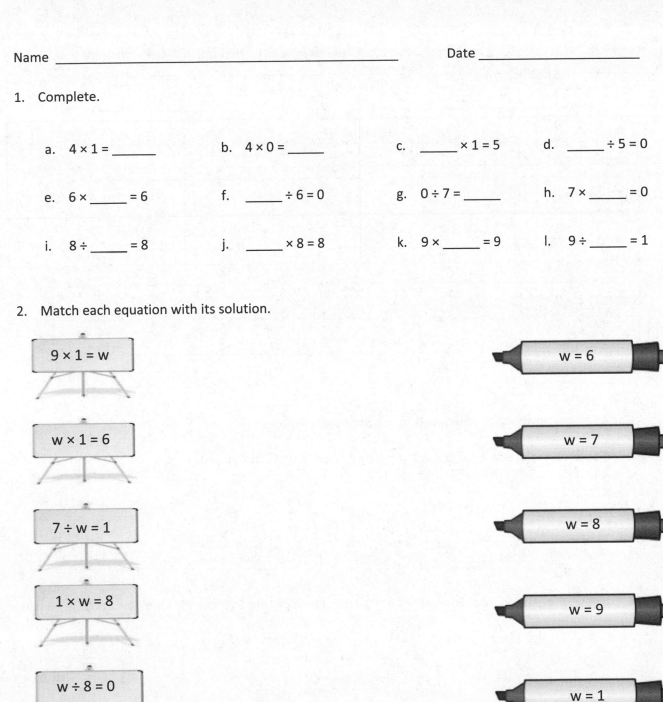

$9 \times 1 = w$	$w = 6$
$w \times 1 = 6$	$w = 7$
$7 \div w = 1$	$w = 8$
$1 \times w = 8$	$w = 9$
$w \div 8 = 0$	$w = 1$
$9 \div 9 = w$	$w = 0$

Lesson 16:   Reason about and explain arithmetic patterns using units
of 0 and 1 as they relate to multiplication and division.

©2015 Great Minds. eureka-math.org
G3-M3-SE-B2-1.3.1-01.2016

77

3. Let $c = 8$. Determine whether the equations are true or false. The first one has been done for you.

a. $c \times 0 = 8$	*False*
b. $0 \times c = 0$	
c. $c \times 1 = 8$	
d. $1 \times c = 8$	
e. $0 \div c = 8$	
f. $8 \div c = 1$	
g. $0 \div c = 0$	
h. $c \div 0 = 8$	

4. Rajan says that any number multiplied by 1 equals that number.

   a. Write a multiplication equation using n to represent Rajan's statement.

   b. Using your equation from Part (a), let $n = 5$, and draw a picture to show that the new equation is true.

Lesson 16: Reason about and explain arithmetic patterns using units of 0 and 1 as they relate to multiplication and division.

©2015 Great Minds. eureka-math.org
G3-M3-SE-B2-1.3.1-01.2016

2. In the table, only the products on the diagonal are shown.

a. Label each product on the diagonal.

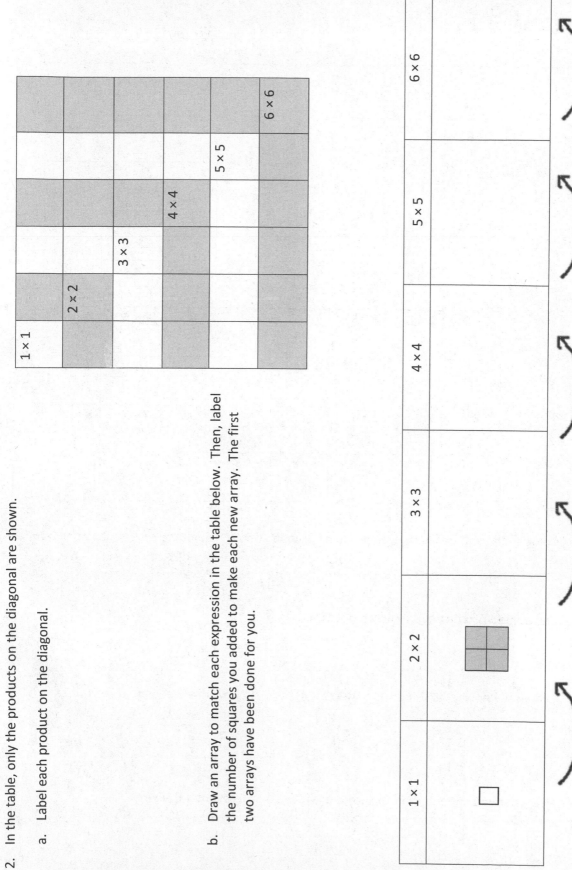

1 × 1					
	2 × 2				
		3 × 3			
			4 × 4		
				5 × 5	
					6 × 6

b. Draw an array to match each expression in the table below. Then, label the number of squares you added to make each new array. The first two arrays have been done for you.

1 × 1	2 × 2	3 × 3	4 × 4	5 × 5	6 × 6
□					

3

Lesson 17: Identify patterns in multiplication and division facts using the multiplication table.

©2015 Great Minds. eureka-math.org
G3-M3-SE-B2-1.3.1-01.2016

Name _____   Date _____

1. Write the products into the squares as fast as you can.

1 × 1	2 × 1	3 × 1	4 × 1	5 × 1	6 × 1	7 × 1	8 × 1
1 × 2	2 × 2	3 × 2	4 × 2	5 × 2	6 × 2	7 × 2	8 × 2
1 × 3	2 × 3	3 × 3	4 × 3	5 × 3	6 × 3	7 × 3	8 × 3
1 × 4	2 × 4	3 × 4	4 × 4	5 × 4	6 × 4	7 × 4	8 × 4
1 × 5	2 × 5	3 × 5	4 × 5	5 × 5	6 × 5	7 × 5	8 × 5
1 × 6	2 × 6	3 × 6	4 × 6	5 × 6	6 × 6	7 × 6	8 × 6
1 × 7	2 × 7	3 × 7	4 × 7	5 × 7	6 × 7	7 × 7	8 × 7
1 × 8	2 × 8	3 × 8	4 × 8	5 × 8	6 × 8	7 × 8	8 × 8

a. Color all the squares with even products orange. Can an even product ever have an odd factor?

b. Can an odd product ever have an even factor?

c. Everyone knows that 7 × 4 = (5 × 4) + (2 × 4). Explain how this is shown in the table.

d. Use what you know to find the product of 7 × 16 or 8 sevens + 8 sevens.

Lesson 17:   Identify patterns in multiplication and division facts using the
multiplication table.

©2015 Great Minds. eureka-math.org
G3-M3-SE-B2-1.3.1-01.2016

7

c. What pattern do you notice in the number of squares that are added to each new array?

d. Use the pattern you discovered in Part (b) to prove this: 9 × 9 is the sum of the first 9 odd numbers.

Name _____ Date _____

1. a. Write the products into the chart as fast as you can.

×	1	2	3	4	5	6	7	8
1								
2								
3								
4								
5								
6								
7								
8								

b. Color the rows and columns with even factors yellow.

c. What do you notice about the factors and products that are left unshaded?

Lesson 17:   Identify patterns in multiplication and division facts using the
             multiplication table.

©2015 Great Minds. eureka-math.org
G3-M3-SE-B2-1.3.1-01.2016

d. Complete the chart by filling in each blank and writing an example for each rule.

Rule	Example
odd times odd equals _____	
even times even equals _____	
even times odd equals _____	

e. Explain how $7 \times 6 = (5 \times 6) + (2 \times 6)$ is shown in the table.

f. Use what you know to find the product of 4×16 or 8 fours + 8 fours.

2. Today in class, we found that $n \times n$ is the sum of the first n odd numbers. Use this pattern to find the value of n for each equation below. The first is done for you.

a. $1 + 3 + 5 = n \times n$

$9 = 3 \times 3$

b. $1 + 3 + 5 + 7 = n \times n$

Lesson 17: Identify patterns in multiplication and division facts using the multiplication table.

83

©2015 Great Minds. eureka-math.org
G3-M3-SE-B2-1.3.1-01.2016

c. $1 + 3 + 5 + 7 + 9 + 11 = n \times n$

d. $1 + 3 + 5 + 7 + 9 + 11 + 13 + 15 = n \times n$

e. $1 + 3 + 5 + 7 + 9 + 11 + 13 + 15 + 17 + 19 = n \times n$

Lesson 17:   Identify patterns in multiplication and division facts using the
              multiplication table.

©2015 Great Minds. eureka-math.org
G3-M3-SE-B2-1.3.1-01.2016

Name _____ Date _____

Use the RDW process for each problem. Explain why your answer is reasonable.

1. Rose has 6 pieces of yarn that are each 9 centimeters long. Sasha gives Rose a piece of yarn. Now, Rose has a total of 81 centimeters of yarn. What is the length of the yarn that Sasha gives Rose?

2. Julio spends 29 minutes doing his spelling homework. He then completes each math problem in 4 minutes. There are 7 math problems. How many minutes does Julio spend on his homework in all?

Lesson 18:   Solve two-step word problems involving all four operations and assess the reasonableness of solutions.

©2015 Great Minds. eureka-math.org
G3-M3-SE-B2-1.3.1-01.2016

85

3. Pearl buys 125 stickers. She gives 53 stickers to her little sister. Pearl then puts 9 stickers on each page of her album. If she uses all of her remaining stickers, on how many pages does Pearl put stickers?

4. Tanner's beaker had 45 milliliters of water in it at first. After each of his friends poured in 8 milliliters, the beaker contained 93 milliliters. How many friends poured water into Tanner's beaker?

5. Cora weighs 4 new, identical pencils and a ruler. The total weight of these items is 55 grams. She weighs the ruler by itself and it weighs 19 grams. How much does each pencil weigh?

Lesson 18:   Solve two-step word problems involving all four operations and assess the reasonableness of solutions.

©2015 Great Minds. eureka-math.org
G3-M3-SE-B2-1.3.1-01.2016

Name _____ Date _____

Use the RDW process for each problem. Explain why your answer is reasonable.

1. Mrs. Portillo's cat weighs 6 kilograms. Her dog weighs 22 kilograms more than her cat. What is the total weight of her cat and dog?

2. Darren spends 39 minutes studying for his science test. He then does 6 chores. Each chore takes him 3 minutes. How many minutes does Darren spend studying and doing chores?

3. Mr. Abbot buys 8 boxes of granola bars for a party. Each box has 9 granola bars. After the party, there are 39 bars left. How many bars were eaten during the party?

EUREKA MATH

Lesson 18:   Solve two-step word problems involving all four operations and assess the reasonableness of solutions.

87

©2015 Great Minds. eureka-math.org
G3-M3-SE-B2-1.3.1-01.2016

4. Leslie weighs her marbles in a jar, and the scale reads 474 grams. The empty jar weighs 439 grams. Each marble weighs 5 grams. How many marbles are in the jar?

5. Sharon uses 72 centimeters of ribbon to wrap gifts. She uses 24 centimeters of her total ribbon to wrap a big gift. She uses the remaining ribbon for 6 small gifts. How much ribbon will she use for each small gift if she uses the same amount on each?

6. Six friends equally share the cost of a gift. They pay $90 and receive $42 in change. How much does each friend pay?

Lesson 18:   Solve two-step word problems involving all four operations and assess the reasonableness of solutions.

Name _____   Date _____

1. Use the disks to fill in the blanks in the equations.

a.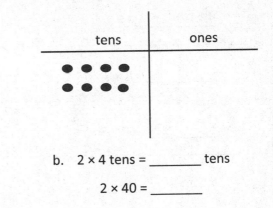

4×3 ones = _____ ones

$4 \times 3 =$ _____

b.

4×3 tens = _____ tens

$4 \times 30 =$ _____

2. Use the chart to complete the blanks in the equations.

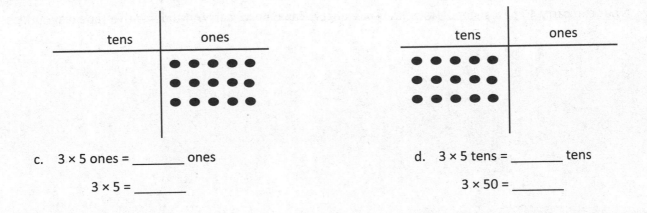

tens	ones

a. 2×4 ones = _____ ones

$2 \times 4 =$ _____

tens	ones

b. 2×4 tens = _____ tens

$2 \times 40 =$ _____

tens	ones

c. 3×5 ones = _____ ones

$3 \times 5 =$ _____

tens	ones

d. 3×5 tens = _____ tens

$3 \times 50 =$ _____

EUREKA MATH™

Lesson 19:   Multiply by multiples of 10 using the place value chart.

89

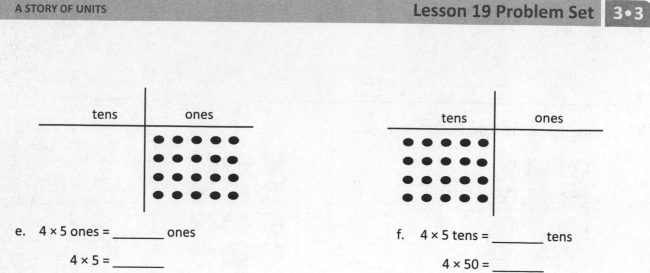

e.   4 × 5 ones = _____ ones

     4 × 5 = _____

f.   4 × 5 tens = _____ tens

     4 × 50 = _____

3. Fill in the blank to make the equation true.

a. _____ = 7 × 2	b. _____ tens = 7 tens × 2
c. _____ = 8 × 3	d. _____ tens = 8 tens × 3
e. _____ = 60 × 5	f. _____ = 4 × 80
g. 7 × 40 = _____	h. 50 × 8 = _____

4. A bus can carry 40 passengers. How many passengers can 6 buses carry? Model with a tape diagram.

Lesson 19:   Multiply by multiples of 10 using the place value chart.

©2015 Great Minds. eureka-math.org
G3-M3-SE-B2-1.3.1-01.2016

Name _____   Date _____

1. Use the disks to complete the blanks in the equations.

a.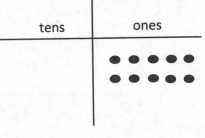

$$3 \times 3 \text{ ones} = \underline{\hspace{1.5cm}} \text{ ones}$$

$$3 \times 3 = \underline{\hspace{1.5cm}}$$

b.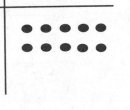

$$3 \times 3 \text{ tens} = \underline{\hspace{1.5cm}} \text{ tens}$$

$$30 \times 3 = \underline{\hspace{1.5cm}}$$

2. Use the chart to complete the blanks in the equations.

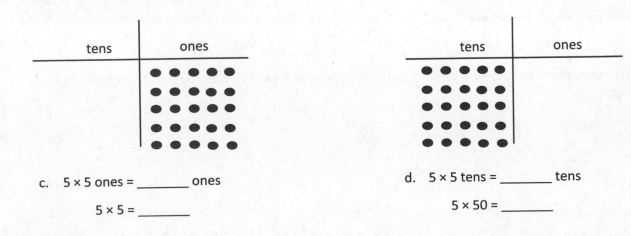

a. $2 \times 5 \text{ ones} = \underline{\hspace{1.5cm}} \text{ ones}$

$$2 \times 5 = \underline{\hspace{1.5cm}}$$

b. $2 \times 5 \text{ tens} = \underline{\hspace{1.5cm}} \text{ tens}$

$$2 \times 50 = \underline{\hspace{1.5cm}}$$

c. $5 \times 5 \text{ ones} = \underline{\hspace{1.5cm}} \text{ ones}$

$$5 \times 5 = \underline{\hspace{1.5cm}}$$

d. $5 \times 5 \text{ tens} = \underline{\hspace{1.5cm}} \text{ tens}$

$$5 \times 50 = \underline{\hspace{1.5cm}}$$

EUREKA
MATH™

Lesson 19:   Multiply by multiples of 10 using the place value chart.

91

©2015 Great Minds. eureka-math.org
G3-M3-SE-B2-1.3.1-01.2016

3. Match.

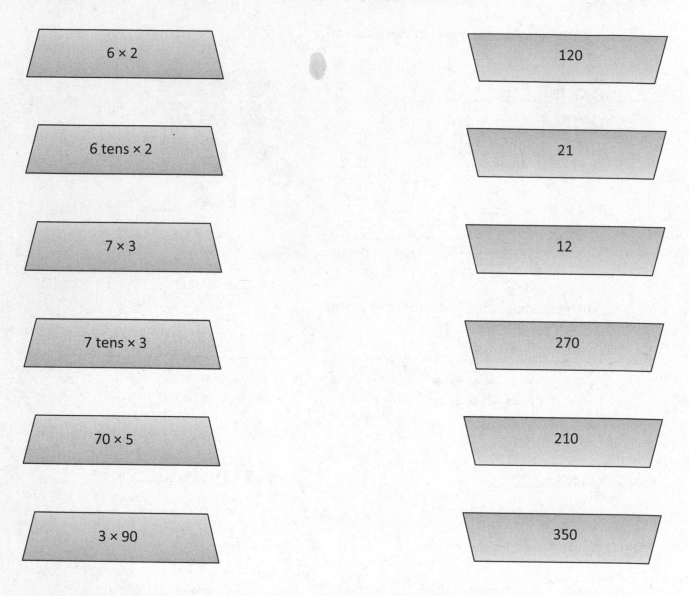

6 × 2	120
6 tens × 2	21
7 × 3	12
7 tens × 3	270
70 × 5	210
3 × 90	350

4. Each classroom has 30 desks. What is the total number of desks in 8 classrooms? Model with a tape diagram.

Lesson 19:   Multiply by multiples of 10 using the place value chart.

EUREKA MATH

Name _____   Date _____

1. Use the chart to complete the equations. Then, solve. The first one has been done for you.

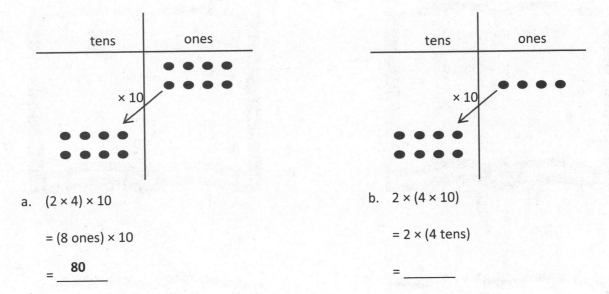

a. (2 × 4) × 10

   = (8 ones) × 10

   = __**80**__

b. 2 × (4 × 10)

   = 2 × (4 tens)

   = _____

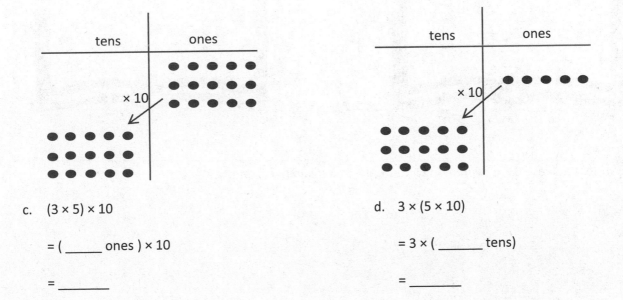

c. (3 × 5) × 10

   = (_____ ones) × 10

   = _____

d. 3 × (5 × 10)

   = 3 × (_____ tens)

   = _____

2. Place parentheses in the equations to find the related fact. Then, solve. The first one has been done for you.

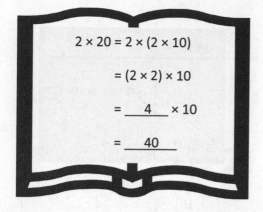

$$2 \times 20 = 2 \times (2 \times 10)$$
$$= (2 \times 2) \times 10$$
$$= \underline{\quad 4 \quad} \times 10$$
$$= \underline{\quad 40 \quad}$$

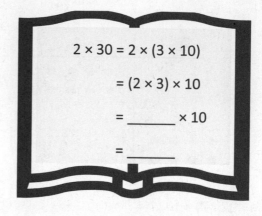

$$2 \times 30 = 2 \times (3 \times 10)$$
$$= (2 \times 3) \times 10$$
$$= \underline{\qquad} \times 10$$
$$= \underline{\qquad}$$

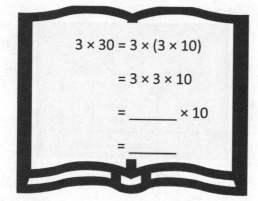

$$3 \times 30 = 3 \times (3 \times 10)$$
$$= 3 \times 3 \times 10$$
$$= \underline{\qquad} \times 10$$
$$= \underline{\qquad}$$

$$2 \times 50 = 2 \times 5 \times 10$$
$$= 2 \times 5 \times 10$$
$$= \underline{\qquad} \times 10$$
$$= \underline{\qquad}$$

3. Gabriella solves 20×4 by thinking about 10×8. Explain her strategy.

Lesson 20:   Use place value strategies and the associative property
$n \times (m \times 10) = (n \times m) \times 10$ (where n and m are less than 10) to
multiply by multiples of 10.

©2015 Great Minds. eureka-math.org
G3-M3-SE-B2-1.3.1-01.2016

EUREKA MATH™

Name _____   Date _____

1. Use the chart to complete the equations. Then, solve.

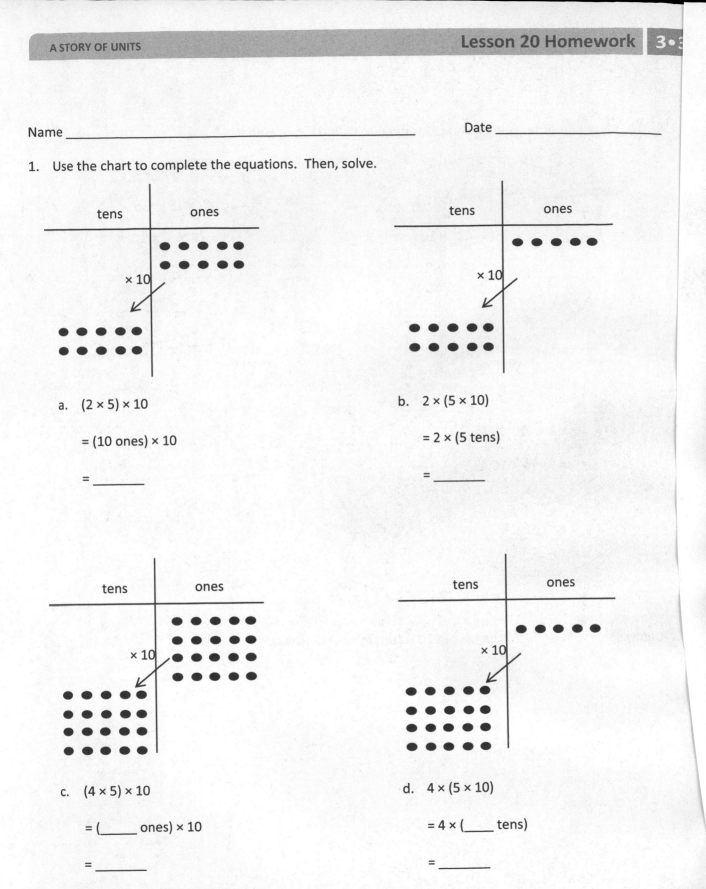

a. $(2 \times 5) \times 10$

= (10 ones) × 10

= _____

b. $2 \times (5 \times 10)$

= 2 × (5 tens)

= _____

c. $(4 \times 5) \times 10$

= (_____ ones) × 10

= _____

d. $4 \times (5 \times 10)$

= 4 × (_____ tens)

= _____

Lesson 20:   Use place value strategies and the associative property
$n \times (m \times 10) = (n \times m) \times 10$ (where n and m are less than 10) to
multiply by multiples of 10.

©2015 Great Minds. eureka-math.org
G3-M3-SE-B2-1.3.1-01.2016

95

2. Solve. Place parentheses in (c) and (d) as needed to find the related fact.

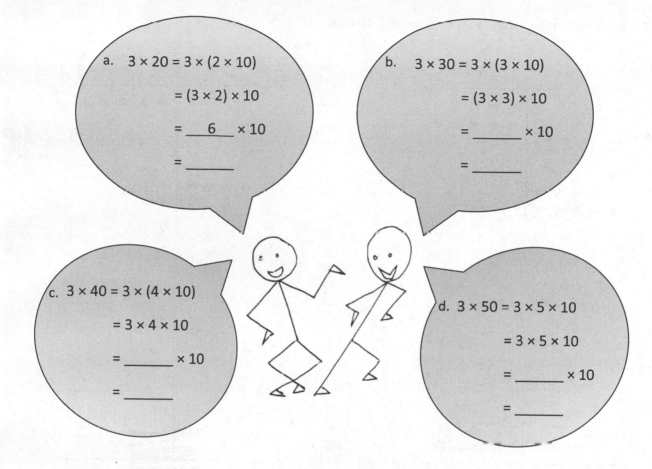

a. $3 \times 20 = 3 \times (2 \times 10)$

$= (3 \times 2) \times 10$

$= \underline{6} \times 10$

$= \underline{}$

b. $3 \times 30 = 3 \times (3 \times 10)$

$= (3 \times 3) \times 10$

$= \underline{} \times 10$

$= \underline{}$

c. $3 \times 40 = 3 \times (4 \times 10)$

$= 3 \times 4 \times 10$

$= \underline{} \times 10$

$= \underline{}$

d. $3 \times 50 = 3 \times 5 \times 10$

$= 3 \times 5 \times 10$

$= \underline{} \times 10$

$= \underline{}$

3. Danny solves 5×20 by thinking about 10×10. Explain his strategy.

Lesson 20:   Use place value strategies and the associative property
$n \times (m \times 10) = (n \times m) \times 10$ (where n and m are less than 10) to
multiply by multiples of 10.

©2015 Great Minds. eureka-math.org
G3-M3-SE-B2-1.3.1-01.2016

Name _____   Date _____

Use the RDW process to solve each problem. Use a letter to represent the unknown.

1. There are 60 seconds in 1 minute. Use a tape diagram to find the total number of seconds in 5 minutes and 45 seconds.

2. Lupe saves $30 each month for 4 months. Does she have enough money to buy the art supplies below? Explain why or why not.

Art Supplies
$142

3. Brad receives 5 cents for each can or bottle he recycles. How many cents does Brad earn if he recycles 48 cans and 32 bottles?

4. A box of 10 markers weighs 105 grams. If the empty box weighs 15 grams, how much does each marker weigh?

5. Mr. Perez buys 3 sets of cards. Each set comes with 18 striped cards and 12 polka dot cards. He uses 49 cards. How many cards does he have left?

6. Ezra earns $9 an hour working at a book store. She works for 7 hours each day on Mondays and Wednesdays. How much does Ezra earn each week?

Lesson 21:   Solve two-step word problems involving multiplying single-digit factors and multiples of 10.

©2015 Great Minds. eureka-math.org
G3-M3-SE-B2-1.3.1-01.2016

Name _____ Date _____

Use the RDW process for each problem. Use a letter to represent the unknown.

1. There are 60 minutes in 1 hour. Use a tape diagram to find the total number of minutes in 6 hours and 15 minutes.

2. Ms. Lemus buys 7 boxes of snacks. Each box has 12 packets of fruit snacks and 18 packets of cashews. How many snack packets does she buy altogether?

3. Tamara wants to buy a tablet that costs $437. She saves $50 a month for 9 months. Does she have enough money to buy the tablet? Explain why or why not.

Lesson 21: Solve two-step word problems involving multiplying single-digit factors and multiples of 10.

©2015 Great Minds. eureka-math.org
G3-M3-SE-B2-1.3.1-01.2016

99

4. Mr. Ramirez receives 4 sets of books. Each set has 16 fiction books and 14 nonfiction books. He puts 97 books in his library and donates the rest. How many books does he donate?

5. Celia sells calendars for a fundraiser. Each calendar costs $9. She sells 16 calendars to her family members and 14 calendars to the people in her neighborhood. Her goal is to earn $300. Does Celia reach her goal? Explain your answer.

6. The video store sells science and history movies for $5 each. How much money does the video store make if it sells 33 science movies and 57 history movies?

Lesson 21:   Solve two-step word problems involving multiplying single-digit factors and multiples of 10.

©2015 Great Minds. eureka-math.org
G3-M3-SE-B2-1.3.1-01.2016